# COSMOLOGY, ECOLOGY, AND THE ENERGY OF GOD

# Cosmology, Ecology, and the Energy of God

*Edited by*

DONNA BOWMAN AND CLAYTON CROCKETT

FORDHAM UNIVERSITY PRESS

*New York* 2012

Library of Congress Cataloging-in-Publication Data

Interdisciplinary Conference on Theology and Energy
(2009 : University of Central Arkansas and Hendrix
College) Cosmology, ecology, and the energy of God /
edited by Donna Bowman and Clayton Crockett. — 1st ed.
p. cm.
Includes bibliographical references and index.
ISBN 978-0-8232-3895-8 (cloth : alk. paper) —
ISBN 978-0-8232-3896-5 (pbk. : alk. paper)
1. Nature—Religious aspects—Congresses.
2. Ecotheology—Congresses. 3. Power resources—
Miscellanea—Congresses. 4. Cosmology—Congresses.
I. Bowman, Donna. II. Crockett, Clayton, 1969– III. Title.
BL65.N35157 2009
201'.77—dc23
2011026673

Printed in the United States of America
14 13 12   5 4 3 2 1
First edition

*For Norb*

# CONTENTS

     Bringing Earth Back to Heaven
         LUKE B. HIGGINS                                                         121

9.   "Go Big or Go Home": A Critique of the Western
     Concept of Energy/Power and a Theological Alternative
         OZ LORENTZEN                                                            141

10.  God Is Green; or, A New Theology of Indulgence
         JEFFREY W. ROBBINS                                                      151


     *Notes*                                                                     163
     *List of Contributors*                                                      191
     *Index of Names and Titles*                                                 195

# ACKNOWLEDGMENTS

The editors thank Helen Tartar at Fordham University Press for her interest in and support of this project, as well as two encouraging and helpful readers for the press. This volume emerged from the Interdisciplinary Conference on Theology and Energy (ICTE) held in February 2009 at the University of Central Arkansas (UCA) and Hendrix College. We thank the sponsors of this conference, including UCA Honors College, UCA Department of Philosophy and Religion, UCA College of Liberal Arts, UCA Humanities and World Cultures Institute, Hendrix College Steel Center for the Study of Religion, and the American Academy of Religion. We would also like to thank the participants and attendees of this conference, including the contributors to this book, as well as Don Viney, Kevin Mequet, and Andrew Saldino.

# Cosmology, Ecology, and the Energy of God

# Introduction

*Donna Bowman and Clayton Crockett*

What does energy have to do with theology? We face an energy crisis, an energy deficit, not only in material resources but also in terms of creative thinking. Theology is reflection about what concerns us ultimately, and many theologians and religious people give that object of ultimate concern the name God. Today we live in a world that is threatened in ecological and environmental ways, and many humans, including theologians, are drawing on traditional, modern, technical, and spiritual resources to reflect upon and provide resources to intervene with this situation. Eco-theology is a vital and important discourse, but there has been almost no explicit attention to the topic of energy from a theological perspective. One counterinstance, however, is Flora Keshgegian, who, toward the end of her book *God Reflected: Metaphors for Life*, invokes the metaphor of energy. She asks: "What happens to our image of God if the divine is understood to be energy for life? What difference would it make if God were not a being, but energy as the really real that makes life happen?"[1]

Following upon Keshgegian's question, *Cosmology, Ecology, and the Energy of God* brings together a group of essays by creative theologians and theorists of religion on the interdisciplinary topic of energy. The contributors of this volume think across some of these religious, environmental, and scientific boundaries in order to grapple powerfully with the topic of energy in multifaceted ways. We want to briefly mention three aspects of energy that are relevant today and that provoke theoretical and theological reflection: energy resource scarcity, cosmological dark energy, and spiritual energy.

We currently exist in a state of energy crisis due to increasing costs and approaching limits of fossil fuel extraction and production. Even if oil production has not yet peaked globally, it has become more expensive and more difficult to extract, leading to petroleum exploration of tar sands and deeper and deeper on the ocean floor. The April 2010 British Petroleum oil spill in the Gulf of Mexico is a sad result of this situation. The Deepwater Horizon rig was attempting to extract oil at extreme depths, where proven technical methods and reliable safety mechanisms are lacking. The fact that this ecological disaster has resulted in only a temporary six-month moratorium on deepwater drilling despite the devastating environmental and economic damage caused by the spill is a testament to just how desperate we are for oil and how precarious our economy (both in the United States and globally) remains in the wake of the financial crisis of 2008. Furthermore, we recognize that while fossil fuels are extremely cheap sources of stored solar energy, they are finite in quantity and polluting in practice. Energy scarcity is a serious global concern, and one question is whether the physical scarcity mandated by earth's limits necessarily entails metaphysical scarcity at the level of creative ideas.

On another plane, cosmological physicists discovered in 1998 that the universe's expansion appears to be accelerating, and they posit something they call dark energy as the driver of that acceleration. Dark energy is "the extra ingredient that cosmologists require in order to balance the density of the universe and account for why its expansion is accelerating."[2] This balancing act is extraordinary, because whatever dark energy is, it seems to make up more than three-quarters of the matter/energy of the universe. Dark energy requires a kind of cosmological constant, whether it is actually constant or in fact declining ever so slightly, to balance the observed acceleration of the universe's expansion with what we know about its content, including the four basic forces, and about light and dark matter. Dark energy currently represents a frontier and a limit of experimental and theoretical understanding, and even though this frontier is constantly shifting, it still raises metaphysical and theological questions about the nature and fate of the universe. What we know about the universe we live in may be much more unsettled at the edges than many nonscientists realize.

Finally, spiritual energy has long been a traditional topic of interest for some of the most creative religious and philosophical thinkers. How do we connect spiritual and metaphysical conceptions of energy, including divine energy, with physical and material notions of energy? We suggest that just as Einstein's theory of relativity implies the breakdown of a hard and fast distinction between matter and energy, it makes sense to view energy as material and spiritual at the same time rather than to dualistically oppose them. Energy, as Keshgegian suggests, provides a vital metaphor with which to conceptualize the divine, and at the same time, energy is vitally important to what we do in practical, political, and environmental ways.

Given how traditional theological reflection has often been divorced from the scientific understanding of the natural world, this volume is intended as a timely and urgent intervention by advancing a more open theological approach that does not simply assume the framework of orthodox Christianity or any particular denominational perspective from which to ask theological questions about energy. We seek at least a provisional agnosticism in theology, to draw from the title and topic of Whitney Bauman's essay, but this agnosticism is a healthy acknowledgment of uncertainty rather than an apathetic response to a complex situation.

In terms of methodology, two theological perspectives are privileged, which we think can fruitfully work together to form a hybrid mode of generative theological creativity. Following the example of Catherine Keller, who pioneers this model of working with and between these two perspectives, the contributors to this book draw on process and postmodern deconstructive models of theological thought that provide an important critical intersection not only between the topic of theology and energy but also between these two distinct schools or traditions of thought.

American postmodern theology emerged out of the death-of-God theologies of the 1960s, which linked up with insights derived from French poststructuralism and deconstruction. Here theological reflection is cut free from orthodoxy and confessional obligations and becomes a tool to ask important questions about language, culture, and the production of meaning. However, postmodern theology followed the linguistic turn and failed to pay much attention to science and nature. In this way, process thought provides an important supplement and corrective to it, without rejecting the significance of postmodern epistemological and cultural insights. At the same time, postmodern theologians have more recently taken up more explicitly political issues and questions, following the later work of Jacques Derrida, as well as Continental philosophers such as Slavoj Žižek, Antonio Negri, Alain Badiou and Giorgio Agamben. This political turn opens up on issues of environment, ecology, energy, and distribution of resources and wealth.

Process theology, inspired by the philosophy of Alfred North Whitehead and formulated by John Cobb and others, has been a theological tradition that has most explicitly engaged with scientific and environmental understandings of the natural world and has attempted to reconcile theological understandings of God with nature. At times, however, process theology has been inhibited by the metaphysical straitjacket of Whitehead's extraordinary conceptual edifice and forced to defend this edifice rather than risk theologically becoming otherwise. Gilles Deleuze, a French poststructuralist philosopher, has engaged Whitehead's work constructively and creatively, and this cross-pollination has encouraged other process and postmodern theologically inclined thinkers, most notably Catherine Keller, to follow this lead. Keller's book *Face of the Deep: A Theology of Becoming* is one of the most profound works of constructive theological reflection of this century. Keller marshals the resources of process theology, postmodernism, feminism, postcolonialism, and biblical criticism to develop a sustained critique of the traditional theological doctrine of *creatio ex nihilo*, and instead sees creation as emerging from a profound depth that manifests provocatively along the edge of chaos.

This collection features an essay by Keller herself, as well as other process and postmodern thinkers as they engage with the topic of energy in creative theological ways. *Cosmology, Ecology, and the Energy of God* emerges, then, from two distinct discourses moving toward each other—on the one hand, discussions and books about environmental or ecological theology from a process theological perspective, as represented in the work of Keller and Jay McDaniel, among others. On the other hand, this book links with discussions from a more postmodern approach to religion and theology that are increasingly taking up political and economic issues, represented here by Jeffrey W. Robbins and Clayton Crockett.

The essays in this volume make a strong case that it is impossible to understand, affect, or plan for our utilization of energy in the future without taking into account energy's theological dimension. Theological thought brings to the discussion, currently dominated by consideration of energy's economic and ecological value, perspectives on energy's aesthetic, moral, and spiritual values. Our contributors exhibit a deep appreciation for the indispensibility of energy in living systems, which brings with it the imperative to conceptualize energy in a positive light. Many of the authors are critical of the reductionist ways energy has been treated and talked about since the advent of the industrial age. But many are also hopeful that tools for a more complete discourse are at hand, and point to models and organizations that demonstrate promising approaches to energy. It is also worth noting the creativity inherent in the diverse literary structures; disciplinary dialogue partners; and historical, philosophical, and theological reference points on display in this collection. In these essays, as in the message this book attempts to convey, the vitality required to find a new way forward is structurally demonstrated.

Catherine Keller's opening essay seasons a process theological approach with literature, ecology, physics, and liturgy for a "transdisciplinary" survey of the topic. More importantly, her playful tone evokes the "delight" mentioned in her epigraph (from William Blake) and creates a space in which ideas from many sources can be utilized for their suggestive power. Energy, after all, is not one category among many; it is *the* category into which all others may be transformed and in light of which many divisive distinctions may disappear. To summon the energy to confront our energy problems often seems taxing or impossible, yet Keller gives us a way of reconceptualizing the scope of energy—and its theological dimensions in the sheer facticity of the immense universe, quite apart from any particular doctrine of God—that reopens its flow through us for our creative engagement with the work we have been given to do.

How might a shift in thinking from the industrial to the metaphysical alter our perspective on energy? Mary-Jane Rubinstein channels Keller's palpable sense of delight in her exploration of "dark energy," that potent and poignant scientific concept whose mysteries seem so amenable to theological play. Weaving a narrative with historical, literary, and existential modalities, Rubenstein draws parallels between the energy of divine spirit and the ties that bind (or engine that drives) the whirling galaxies. T. Wilson Dickinson draws a connection between the notion of light (or "solar energy" broadly conceived) and our ability to theorize and poeticize the creation in which we take part. Conversing with Gregory of Nyssa, Dickinson advocates for theology and philosophy as an "alternative energy" source to balance the overwhelmingly scientific, quantified, and data-processed notions of energy that dominate the cultural discussion.

For those of us who came of age in the twentieth century, the new complexity, fluidity, and interdependence of science's emerging picture of energy can be startling and strange. Clayton Crockett considers the reductionistic concept of energy as a thermal measure of particle movement and suggests that this definition, when coupled with theologies that locate value in a transcendent realm and correspondingly tend to denigrate human existence on earth, provides an explanation for the history of energy exploitation. Using the architecturally based speculative work of Kevin Mequet, Crockett proposes a non-thermodynamic framework for understanding energy and relates it to a mature materialist theology that locates value in the energetic transformations of the immanent plane. Whitney Bauman contrasts the Newtonian concept of equilibrium with a more complete and open nonequilibrium thermodynamics to urge us toward nonsystematic and nonlinear ways of thinking in both science and the humanities. Breaking free from foundational methodologies, Bauman argues, moves us toward less rigid epistemologies, perhaps ending at an agnostic stance that opens us up to creative energy and flow.

Moving from the dialogue with science to an engagement with culture, Jay McDaniel resituates the discussion in the context of a set of principles for thought and action based on his experience in China, locus of the development trend that will set the stage for this century's sustainability prospects. In a post-American world, McDaniel argues, the countercultural theological and philosophical stream of Whiteheadian thought can give us traction in our efforts to chart a new course. McDaniel's expansive yet concretely rooted argument moves us toward constructive proposals grounded in emerging scientific consensus and informed by the most promising contemporary theological movements. Donna Bowman finds a model for channeling and fostering the energy of personal production in a number of crowd-sourced social networks. She connects the Web 2.0 revolution to the flow of creativity as envisioned by process theology, arguing for a recovery of the notion of divine-human likeness in the activity of making, filtered through the shared and amplified energy of community.

Luke B. Higgins turns our attention to the conceptual roots of the present energy and ecological crises with a judicious application of Alfred North Whitehead's famous "fallacy of misplaced concreteness." By treating energy's exchange and work-potential value as the locus of reality in a wide variety of natural and physical processes, modern science has bequeathed to the societies dependent on it a "bifurcated" worldview: creative humans turning otherwise wasted energy into value through industrial processes. Higgins offers an alternative concept of "alliance" with the natural world to repair the rift and reclaim the energy in operation all around us as already valuable. Oz Lorentzen offers a personalistic approach to the topic of energy, moving away from the materialistic categories and interactions that have governed scientific discourse about energy. In doing so, he moves decisively into the realm of Christian theology by positing theism and incarnationalism as the structuring methodologies for his personalistic proposal. Without knowing the aim of human life, Lorentzen argues, it is impossible to make meaningful claims about well-being both for human populations and for the environmental processes on which they depend.

Finally, Jeffrey W. Robbins calls our attention to three moments in recent history—the American Indian movement, the 1970s oil crisis, and Richard Nixon's abandonment of the gold standard—to make the case that a new religious language is needed to do justice to the new post-petroleum reality upon us. His essay raises the crucial question of our readiness for the spiritual, economic, and cultural transformation that would be needed to enact a new relationship to energy.

# PART I

# Engaging Physical Sciences

# CHAPTER I

# The Energy We Are:
# A Meditation in Seven Pulsations

## Catherine Keller

*Energy is eternal delight.*

WILLIAM BLAKE, "Marriage of Heaven and Hell"

Delight has the bodily feeling of buoyancy, of dancing lightly, effortless motion, radiant pulsation. The light of delight vibrates, its energy throbs. No wonder modernity turned into an energy addict. Yet the manic excitations that rock our civilization—instant communication, zippy cars, frequent flights, processed foods, monstrous war waste, even the air conditioning that corrects natural summer slowdown—also drive our dependence on nonrenewable energy. It would seem that we have arrived, we earthlings, at a great turning point. Either we take responsibility for our energies—for the planetary effects of our overuse, overproduction, and abundant waste of energy—or we blast our way toward planetary burnout. But how shall we *energize* that sense of responsibility, which may simply feel burdensome, menacing, or paralyzing? To merely speak of energy as a global problem, to objectify energy as a quantifiable substance, begins to drain delight away.

Pin *energy* down to a lexical unit and you get the standard, dull definition: energy is "what does work." But then *work* is defined as "the transfer of energy." A bit circular, no? The closed circle itself seems to drain vitality from the open question of planetary life. Might we instead articulate the open circulation of energy, where the multidimensional and conflicted signifiers of life force and of fuel, of micro- and macrocosmic vibratory fields, even of spiritual potencies and theological metaphors come into play? Might energy do better work for us when we realize that it *is* us?

If energy is most basically not something that we *have* but that we *are*, we might try, experimentally, to think about it differently. We might imagine energy not as a commodity to use nor as a power to calculate, but—with possibly more fidelity both to science and to the spirit of our shared life—as the *rhythm of interactivity*. From this perspective, energy "does work" but never as a mere means to other ends. Its interactions at every level involve worth for their own sake. Energy signifies the pulsation of life: life as a boundless vitality, life that exceeds the distinction between organic and inorganic. In theological terms, energy connotes not only the efforts of work but the effortlessness of grace as well.

In this essay I follow the earthly flow of energy through a rapid series of intersecting contexts, like seven different pulsations. I touch upon each briefly, as in a liturgical circulation. These contexts are (1) an intensification of the question of energy as radically, and even theologically, transdisciplinary; (2) a Blakean sense of the body of energy; (3) our planetary energy policy in its empirical and ecological factuality; (4) eco-process theology and Whitehead's cosmology of emotive energies; (5) a clue about energy in its quantum pulses vis-à-vis the physics of entanglement; (6) a hint of dark energy; and (7) the sacred body of that which "energizes all in all."

## Transdisciplinary Energies

If energy is eternal delight, the spiritual intensity of this volume's question—the question of energy itself—is manifest. Marrying "heaven and hell" in its honor, William Blake meant to mock and subvert the regnant orthodoxy—and, therefore, to provoke theological engagement. Of course, Blake (1757–1828) didn't have to worry about energy policy and ecological catastrophe (though he did protest "the satanic mills" of early modern industrialization). We who do, however, may require a hint of his ferocious delight to fire up our efforts.

Let me suggest that we cannot even get our minds around the elusive notion of "energy" without a spiritually charged sense of its *universe*. But in this wider circulation of meanings, cosmological metaphors slip imperceptibly into the theological. Whether or not you "believe in a personal God," your personal relationship to the unfathomable scale and subtlety of the universe—a universe that is at base not matter, but energy—requires images. Yet this universe itself can hardly be imagined. Not surprisingly, the effort stimulates theological questions, associations, and icons of an all-enfolding relation.

Yet theology (as Feuerbach analyzed it and Blake mocked it) often seems to suck energy up and away from our earthly and embodied vitalities. Furthermore, theology as a self-critical discipline depends upon old-line Christian churches whose energies have been steadily draining away. Thoughtful spirituality in the theological tradition has been enervated both by a disturbingly energetic religiosity on its right flank and scientific materialism on its left. If at the dawn of modernity science was forced to kowtow to the church, by now we have been through a couple centuries of reversal. Here at the end of modernity, the zero-sum game that persists between religion and science obstructs the shared planetary work that might otherwise heal and delight. Fortunately, at certain edges of both theology and science (many of which appear in this volume) new and vital interactions are coming into play.

Might the notion of energy, with its odd mix of literalism and metaphor, science and spirit, fuel a transdisciplinary *feeling* for our planetary collective? By minding the multivalent meanings of energy, might we help to energize a wider responsiveness to the earth, to all of us earthlings? *Earthling*, after all, is a precise translation of *Adam*, from the Hebrew *adamah*, meaning earth, clay, the humus of the human. What if we began to feel the earth and our own "earthbodies" as energetic loci of the universe rather than bodies in commodifiable clods of manipulable matter?[1] Feeling the pulsation of our bodies in our planet, and the pulsation of the planet in its universe, our earthly interactivities are rendered simultaneously intimate and virtually infinite. Perhaps such vibrancy of feeling will help to energize our efforts on behalf of the just and sustainable future that we desire—that we need and must, together, create.

## *The Body of Energy*

If William Blake had to put certain propositions in the mouth of the devil, it was his ironic device for countering the unquestionable absolutes of the Christianity of his day, founded on the following succinctly summarized presupposition of dualism:

> That Man has two real existing principles: Viz: a Body & a Soul.
> That Energy, call'd Evil, is alone from the Body; & that Reason, call'd Good, is alone from the Soul.[2]

Blake was deconstructing the binary of a rational soul divided from the vital body. For the Christian soul, like the Christian God, was assumed to be pure reason, by classical definition devoid of passion or compassion. God, as the unmoved mover, was doctrinally understood to be unmovable, without pathos—apathetic. The earthly turbulence of energy, or potentiality, was readily identified with sensuality, sex, desire—evil. So it took a diabolical parody like that of Blake to expose the demonization of energy. And to propose, with the same didactic "prosiness," a reversal stunning for his epoch:

> Man has no Body distinct from his Soul for that call'd Body is a portion of Soul discern'd by the five Senses, the chief inlets of Soul in this age.
> Energy is the only life and is from the Body and Reason is the bound or outward circumference of Energy.
> Energy is Eternal Delight.[3]

As for the joyful character of energy, we do know that, conversely, psychologically speaking, a persistent and unaccountable lack of energy may be a symptom of depression or of blocked mourning. I had a concentrated experience of this condition recently, near the time of my stepfather's death. For about one week I had literally too little energy to move, even to read. This odd condition of immobility felt altogether bodily. The energy seemed to be dissipating inwardly, not outwardly, as though through a black hole within. Yet the low energy condition allowed (or forced) me to meditate on the relations comprising my own life, on the complications of unrealized relation obstructing a more fluid grief. During this time the metaphoric resonances of depression with the current world situation struck me as eerie. During this time an energy crisis, provoked by the spike in oil prices, had heralded the plunge of the economy toward depression. One might have hoped that the sluggish economy would allow (or force) us as a nation to meditate on the relations that comprise the life of us all, the life of our planetary body.

Of course Blake was not thinking of how to heave our energy flows from the destructive rationality of the carbon-based economy. But he wouldn't be surprised at our situation. For he was railing not only against the orthodox religious dualism and its disdain for the earthly body but also against a scientific materialism that had already produced its own orthodoxy. He satirizes both the figure of Newton and of "old Nobodaddy"—the daddy in heaven who is no-body at all, who "Man" designed in his image to do work for us, and who, in the name of religion, represses the rhythms of the human body and the pulsations of its desires.[4] Blake diagnoses the deadly European dissociations of our souls from our bodies, of our collective rationality from our creaturely flesh. The same great schism between the dominative subject over subjected others (others made objects by repression or exploitation) is now slicing through our planetary life systems. It is not just threatening the whole future project of human civilization but also depriving us—in the present—of the earthly plenitude that comes from participating in the open flow of planetary energies. Instead we have made energy itself into a calculable means of doing work on a world of objects. In order to commodify that energy, we systematically forget that we *are* energy. We are its actualizations, its embodiments, in its ever shifting forms.

The energy of eternal delight is the alternative to a hedonistic indulgence in bursts of pleasure. Pepsi peppiness and consumer glee are draining, not energizing, the planet. The manic excitations produce depressive effects in a vicious circle that, if undiagnosed, blocks the circulation of our spontaneous interactivity. For our souls live indissociably from these bodies that are folds of the living planet. Its life would be the pulse of its energy. As the Blakean delight it is akin to musical or poetic rhythm, to oceanic ebb and flow, to the pulse of blood and the breath of meditation. So this energy, far from a homogenous linear flow, signifies the creative throb of life itself, beyond the distinction of human and nonhuman.

## Energy Policy

According to the first law of thermodynamics, energy doesn't diminish or increase; you cannot create it or destroy it. It shifts shapes. So this second pulsation has us face certain prosaic facts of the current crisis. It is not that we are running out of energy, even if we are at the point of peak oil. Rather, we are running out of a particular form of it. Economics so far has trumped ecology. Indeed it seems almost impossible to imagine that it will not continue to do so, distracting us with jolts of high-energy agitation until the second law of thermodynamics takes over. The entropy of our economic habits will rule out the possibility of an ecologically sustainable civilization. The toothless 2009 Copenhagen Accord, of which nations agreed to "take note," seems only to confirm this entropy.

What had appeared politically impossible in the first years of the new millennium has failed to become actual. But ecological sanity has become at least more *possible*—quite in Derrida's sense of the "impossible possibility of the im/possible." For there is in place, at least in principle, an agreement on the need to prevent a rise of more than two degrees Celsius in the average global temperature over the coming decades. The "eco" of economics could yet amplify its ecological echo. Despite the inevitability of continued disappointment, we are still capable of conducting a planetary metamorphosis to green energy policies, of shifting our fossil fuel economy to a clean energy economy. According to Robert Pollin there are three interrelated projects that could drive an economically feasible environmentalism, that "green new deal" that had been spoken of so optimistically upon President Barack Obama's election: "dramatically increasing energy efficiency; equally dramatically lowering the cost of supplying energy from such renewable sources as solar, wind and geothermal power; and mandating limits and raising prices on the burning of oil, coal and natural gas."[5] Pollin argued that "spending the same amount of money on building a clean energy economy will create three times more jobs within the United States than would spending on our existing fossil fuel infrastructure."[6]

Along with the political will to retool the economy, of course, the science of global warming will continue to lurch and shift alarmingly. For instance, we must process news like the following: "the Intergovernmental Panel warned that the Arctic's 'late-summer sea ice is projected to disappear almost completely towards the end of the 21st century . . .' But, as the new report by the Public Interest Research Center shows, climate scientists are now predicting the end of late-summer sea ice within three to seven years."[7] Hence the proliferation of sorrowfully adorable images of those doomed polar bears. George Monbiot, the British climate writer, commented grouchily, "Forget the sodding polar bears: This is about all of us." As the ice disappears, the region becomes darker, which means it absorbs more heat. The extra warming caused by disappearing sea ice penetrates (932 miles) inland, covering almost the entire region of continuous permafrost. Arctic permafrost contains twice as much carbon as the entire global atmosphere. The methane it mists "is a potent greenhouse gas, packing 25 times more heating power, molecule for molecule, than carbon dioxide."[8] It remains safe as long as the ground stays frozen. But the melting has begun. Methane gas gushers are now gassing out of some places with such force that they keep the water open in Arctic lakes right through the winter. Indeed as the permafrost thaws, it creates new lakes that emit methane. These are the effects of melting permafrost. The terrifying potential of the methane feedback loops to hasten global warming has not yet (as I write) been incorporated into any of the established global climate models, which were menacing enough.

These are sobering findings. But they needn't immobilize us. They do not imply that we have reached the point of no return. Moreover, the United Kingdom Public Interest Research Centre's 2008 "Climate Safety" report wisely warns against the "illusion of certainty" on *either* side of the argument.[9] Climate is a complex system of complex systems, incapable of linear and absolute predictions, however politically expedient claims to certainty seem for environmental activism. So called climate skeptics cling to any admission of uncertainty as evidence against human-caused warming. Nonetheless, the needed *social* energy of transformation over years, indeed decades, will be better stirred and sustained by an open systems model, drawing on risk assessment theory, with its "risk management" strategies and its probabilistic science, appropriate to the nonlinear interactivities of human and nonhuman social systems. An open systems model would avoid the wasteful cycle of apocalyptic hype and complacent denial. A responsible assessment of risks and probabilities requires its own motivation, its own faith—beyond apocalyptic absolutism and misplaced trust. Here the greening of Christianity, given the economic force of its associated populations, becomes a matter of planetary urgency.

## Energy in Eco-Process

Different theologies will serve different contexts. My colleague Laurel Kearns has analyzed three different streams: creation spirituality, evangelical stewardship faith, and eco-social justice theology. Many of us in this volume inhabit the latter, especially as nested in process theology. The process school of thought arose partly in response to the modern tension between religion and science. It originated eight decades ago in Alfred North Whitehead's interpretation of the new—at the time *very* new— physics. He proposed "the shift from materialism to organism" as the basic idea of physical science.[10] That shift would not only *permit* a reconciliation of religion and science, he argued, but would also *require* it. The metaphor of organism "is the displacement of the notion of static stuff by that of *fluent energy.*"[11] *Process* is another name for that energy. It does not flow in a steady stream but rather in distinct pulses, in unique events of interactivity. Process thought thereby roots all experience—which means all existence, human and not—in a cosmological relationality of interacting energy events.

Process thought evolved in the subsequent decades of the twentieth century as the only major school of theological reflection that practiced, simultaneously, on divergent fronts from the 1960s onward, conversation with the natural sciences and the prophetic warnings about our ecological trends. Whitehead's notion of the relation of the divine to the creation (and of all creatures to one another) was ecological avant la lettre and in advance of the recognized ecological crisis. Here God appears as the environment of the world: all creatures dwell within the divine. This is what is often called *panentheism*. Jay McDaniel, above all, has explicated the ecological implications of process panentheism. "God," he writes, "is not merely anthropocentric; God is eco-centric and each living being has its own unique relationship to 'the cosmic lure.' Dogs respond to the lure of God by barking, cats by purring, fish by swimming, and birds by flying."[12]

The converse is also true. As Roland Faber states in a recent piece, developing an "eco-process theology," the "ecological multiplicity of what we call 'world' is the environment for God, 'in which' God 'lives, moves and has God's being.'" The "uniqueness of God, then, is not God's exception from the ecological process but God's exemplification of it par excellence."[13] I engage Faber here in part because he is a leader among those process thinkers supplementing Whiteheadian thought with Continental philosophy, particularly with that of Deleuze, a philosopher who himself was influenced by Whitehead. That methodological bridge carries considerable weight for the transdisciplinary conversation comprising the present book. Deleuze intensifies the event character of every existence in a polyrhythmic, open-ended "intermezzo" of becoming. To name this postmodern cosmos that he finds so precisely anticipated by Whitehead, Deleuze borrows from James Joyce the term *chaosmos*. Its energetic pulses push toward difference, toward intensity. In the movement of process cosmology, however, theology is not discarded, as with Deleuze. Rather, it becomes potentially liberated from its world-transcending, male-divinizing, and human-centered habits.

Faber himself resists even the more anthropocentric kinds of ecological theology. They emphasize, for example, that for the sake of our survival we must stabilize our carbon production and our climate at a certain level. Such theologies may be right. But the rhetoric may be counterproductive. After all, even the alarming Monbiot, grousing about the iconic polar bear, did not say "it is all about us" but "*it is about all of us*." We should not try, contends Faber, "to preserve any status quo of society and the ecological present of the Earth." This is a significant swerve from the rhetoric of preservation and restoration, and it deserves consideration. He would have us instead "transform the Earth and ourselves toward instabilities of deeper intensities and harmonies of deeper complexities—thereby realizing ever more the non-violent 'circle of Love' that is *the metabolism of*

*God and the World.*[14] So, then, in the face of this ecological and economic imperative—in which circulates a theological metabolism—"we should not 'stop' Global Warming (in preserving an already lost *status quo*)." Rather, we should actively *transform ourselves with the Earth* "by learning to understand the world differently." How? "In a way that allows for the recognition of the 'environmental field' of social-economic structures and reforms, by accepting the functional openness of the future of the developing disequilibrium."[15]

Recall that it was Ilya Prigogine, Nobel Prize laureate in thermodynamics, with the frequent collaboration of the Whiteheadian and Deleuzian philosopher of science Isabelle Stengers, who established that complex systems develop only far from conditions of equilibrium, through those nonlinear processes that unfurl at the edge of chaos.[16] In other words, the Whiteheadian notion of fluent energy has been doing work indeed! It conceptually displaces the static notion of "nature" that the environmental movement, surprisingly, had not inherited. Platonic metaphysics, source of the Christian abstract of the rational soul from the energetic body, permeates Western common sense. As Continental science theorist Bruno Latour (who has become interested in Whitehead through Stengers's efforts, thus effecting another form of the methodological bridge the present volume helps to construct) has argued, nature is "that blend of Greek politics, French Cartesianism and American parks." Nature is something, in other words, that has *"nothing to do with nature."*[17] This lumpy amalgam continues to reproduce the detached and disembodied observer, presiding over his flat world of facts. For the sake of both the democratization of the sciences and the empowerment of political ecology, Latour suggests a shift from "matters of fact" to "matters of concern."

Both at the level of a mobilizing rhetoric and of physical description, such postmodern ecologies may energize an intensification of cosmological perspective. But such postmodern transmutations of solid certainties into open interactivities should not be misread as gleeful revolts against life's givens. It is not a matter of disregarding the hard-won stabilities of the material universe—as of the animal body, as of Middle Climate. These concepts, after all, support the evolution of the metabolic intensities that sustain and delight earthbodies. Sustainability emerges for our planet now as the key to complexity. Mere transient events in a void cannot build up a complex system. Therefore, I would want to supplement Faber's emphasis on novelty with Whitehead's own insistence on a certain balance between the conserving and transforming aspects of the world: "The art of progress is to preserve order amid change, and to preserve change amid order." Yet the privilege of the new remains unmistakable: "the pure conservative is fighting against the essence of the universe."[18]

In other words, the shift to clean energy policies will be more likely to happen as the pragmatic codes of environmentalism are resignified—not discarded, but rhythmically absorbed, into a wider and wilder chaosmos. At the theological level, its energetic throbs will become recognizable as "planetary love."[19] For in this vision the divine that serves as a lure to the process of evolution and complexification is a manifestation of cosmic desire. And inasmuch as we earthlings, on our way from dust to dust, meet that lure with delight, we release a fresh energy of symbiotic creativity. The throbbing eco-process universe brings us back, and then takes us forward to this essay's fifth pulsation—that of quantum cosmology.

## Entangled Energies

As a mathematician and a cosmologist, Whitehead, in the mid-1920s, was theorizing the startling new physics of relativity and the quantum just as they were being formulated. What shocked others, including the physicists, delighted him. The paradigm-busting mystery of the metamorphosis of bits of matter into events of energy remains vivid today. And process thought helps render both the shock and the surprise a resource for planetary love. Despite the theism of his metaphysics, a subset of physicists today (including Henry Stapp, Shimon Malin, and Timothy Eastman) read Whitehead to come to grips with a paradigm shift that has not quite happened. The anthropocentrism of the globo-religio-techno-economy *should* have dissolved into the subtle events of flowing energy but instead is all too well fueled not only by an all-consuming "economism" but also by a stubborn Newtonianism.[20] We still presume a world of things made up of inert bits of matter externally related, down to smooth atoms vacuously enduring—all ready to be manipulated and marketed by our transcendent reason.

Of the fluent energy noted above, Whitehead writes: "Such energy has its structure of action and flow, and is inconceivable apart from such structure. It is also conditioned by 'quantum' requirements."[21] In other words, energy comes in quanta—that is, in packets, or pulses—not in a linear and smooth stream. Existence is made of such events, by which Whitehead means every level of existence, of subject, not just of the quantum wave/particles of physics. So it is in the event of this microcosm, this "actual entity," that the shift to the notion of energy as a pulse of interactive actualization finds its home in philosophy. "Mathematical physics translates the saying of Heraclitus, 'All things flow,' into its own language. It then becomes, 'All things are vectors.'" But this means for Whitehead that "All flow of energy obeys 'quantum' conditions."[22] For the universe is comprised of *events* emerging through their feelings of other events; the past events *there* transfer their energy to *here*. "In the language of science, [this direct perception] describes how the quantitative intensity of localized energy bears in itself the vector marks of its origin, and the specialties of its specific forms." In other words, waves of energy do not convey undifferentiated existence, but vectors of difference. The past is inherited by the immediate subject as a "pulse of experience." That pulse can, "under an abstraction, be conceived as the transference of throbs of emotional energy."[23]

Does this "emotion" bring on its own anthropomorphism? Or, au contraire, does it bring a cosmological redistribution of goods (like subjective experience and feeling) that had been anthropocentrically reserved for humans? At any rate, the nonhuman—all of it, including the divine—is thus endowed with a pathos once reserved for the human. The quantum physicist Shimon Malin has written a book inspired by what he calls the radical "paradigm shift suggested by Whitehead: that the ultimate 'atoms of reality' are experiences!"[24] Reading Whitehead, he comments that "the basic constituent units of the universe are 'throbs of experience.'" How does this idea compare with quantum mechanics? A complete agreement between Whitehead's philosophy and quantum mechanics cannot be expected, because, as Malin, reflecting on Schroedinger's formulation, reminds us, the latter is subject to the principle of objectivation, which excludes experiences from its domain of inquiry. It turns out, however, that elementary quantum events come as close to Whitehead's "throbs of experience" as objectivized entities can. Quantum mechanics, as Malin demonstrates eloquently, "adds credence to the Whiteheadian vision, and Whitehead's system helps us to understand the apparently weird features of the quantum domain."[25]

Quantum weirdness is at its highest pitch in what Einstein called, disparagingly, "spooky action at a distance."[26] However, over the course of many decades, the experiments associated with Bell's theorem have confirmed this spookiness. "Two events that take place at the same time seem to influence each other"—even, as it turns out, at an astronomical distance. Malin offers a succinct account of a mind-boggling phenomenon, one that I suspect will pry open the rim of a new paradigm for the chaosmos. "Even when the events take place very far apart they seem to be 'entangled,' they seem to 'feel' each other. Yet the theory requires that the

influence propagate faster than light, which was for Einstein impossible."
In this impossibility there comes into focus an astonishing challenge to
the classical understandings of space and time as local. It "suggests that
once connected in any elemental sense, two particles—once separated—
actually do not cease to be connected. If you interfere with particle **a**, its
entangled particle **b**—even if it has gotten itself to a galaxy a billion light
years away—will respond as though you have interfered with it too. And
it will respond instantaneously." Malin suggests that "such a connection
takes place because both events form a single creative act, a single 'actual
entity,' arising out of a common [energy] field of potentialities."[27]

News of this mysterious nonlocal entanglement is popping up under
many different guises—as quantum nonseparability (Bernard D'Espagnat)
or the implicate order (David Bohm).[28] It is often discussed with reference
to quantum tunneling and teleportation. Nonlocality might not seem like
the best metaphor for our ecological hope, however. Isn't it precisely the
local—local communities, ecologies, agricultures—that has been violated
by global capitalism? *Locovore* has rightly been enshrined in recent dic-
tionaries! But the local is not as straightforward as it seems, whether for
thought or action. Whitehead warned that our Western, common-sense
materialism is based on a "fallacy of simple location." Environmentalism,
as implied by Faber above, is still curing itself of this fallacy. We realize
that in order to redeem a rich sense of the local in the face of already irre-
versible damage to most localities, indeed in order to save our places, neither thinking nor acting locally will suffice. Every locality—every oxygen
molecule in the atmosphere or every snowflake in the Himalayas—is now
entangled in the endlessly complex eco-economics of our globality: such is
our planetary interactivity.[29] Each local quantum throb of experience is a
node in a net of fluent energies that entangles us in the far reaches of the
planet, which entangle the planet itself in its atmosphere, its atmosphere
in its solar system, in its galaxy, in its universe, its expanse into a multiverse whose magnitudes and multiplicities escape all canons of confident
knowledge.

## Dark Energy

We have learned only recently that the expansion of the universe is accelerating. Hence we have another new shocker in physics: *dark energy*.
It is called dark because it is unknown; the name is a placeholder for the
invisible *something* that seems to drive the acceleration of the supernovas

being measured. It acts in opposition to the gravity that attracts them. And it seems to make up more than two-thirds of what exists. But the mystery deepens. As physicist Lee Smolin puts it, "most kinds of matter are under pressure, but the dark energy is under tension—that is, it pulls things together rather than pushes them apart." Nonetheless, "it causes the universe to expand faster. If you are confused by this," he adds kindly, "I sympathize."[30]

In other words, this is so attractive a force that it has the opposite effect: it *attracts* so powerfully that it *repels*. (You know people like this?) Or it is such an attractive force that rather than slowing the expansion down, it has (in general relativity) the effect of accelerating it. British astronomer Percy Seymour proposes a connection between the dark energy and quantum entanglement. He suggests that the quantum waves form a world-line web, and "the cosmic evolution and recycling of matter means . . . that the whole of space is crisscrossed by the web."[31] One returns to Whitehead: "everything is in a certain sense everywhere at all times. For every location involves an aspect of itself in every other location."[32] In the early universe, Seymour writes, "the decay of neutrons to form protons, electrons and neutrinos gave rise to myriads of thin bundles of electric lines of force encased in insulating space, and because of the turbulence then present, these bundles became very entangled. . . . Dark matter comes from the energy within these stretched sheaths."[33] So this intimate entanglement becomes the source for dark energy when the tension reaches cosmological dimensions.

These dimensions immediately escape the purview of this essay. But I do want to note that the theory of the accelerating or "runaway universe" conveys to many a sense of increasingly lonely disconnection in the dark void of an inhuman cosmos. This may just be a projection of our late modern alienation, echoing premodern associations of darkness with the diabolical. But a cosmological "marriage of heaven and hell" may be happening at the creative edges of natural science. We are learning of an immediate connectivity operating across the widest distances, where there is no empty void but rather an infinitely plastic body of mysterious energy. And the very energy of the expansion may flow from the intimacy of the entanglement. Never mind the math. Consider the metaphor! The ancient mystical trope of the "brilliant darkness"—the glowing darkness of the infinite whom we have nicknamed God—seems to be growing (in theory) a subtle body.[34] A body of energy, no thing, but pulsing webs, strings, and fields, the entangled intensity of everything that is, in some sense, everywhere.

## *Energizing All in All*

Does the physicality of energy in postmodern cosmology not appear to be worthy, in a way that vacuous, inert modern matter was not, of the metaphor of the universe as "the body of God"?[35] A body woven of subtle intercommunicating energies and pulsing with a jazzy, multilayered life, it seems to be growing and yet not dissipating, or dissociating, one part of itself from any other. At any rate, these mysteries of physics seem to be portents of a future cosmology in which religion and science can be in cahoots in ways that no longer sever the subject from the object, the mind from the body, the spirit from the universe, and our values from our facts. However, the symbiosis of God and world forms not a unified One but a boundless multiplicity. Enfolded in the manifold of finite differences, the specific limits of our earthly ecologies contour our creativity, give it shape and difference.

I, for one, can no longer stir up enough energy for theology without returning to that dark brilliance of unknowing. This is not just a matter of theological representation. Across the fields, from quantum uncertainty, to philosophical deconstruction, to Judith Butler's relational "opacity," a certain unknowingness increasingly appears as the precondition of any knowledge worth knowing. Launched without mindfulness of their own radical limits, metaphors either transmute energy to aggression or limply dissipate. Yet negative theology—in, for example, its hermeneutics of bib-lical metaphor—also yields creative affirmations. The Genesis figure of the "darkness on the face of the deep" glows with the light that has not yet been separated out of it (as it will be on the first of the seven symbolic days). Like the womb of God in Job, from which the waves of the world come forth, it never closes.[36] Creation at this level, as I have interpreted it elsewhere at length, is captured less by the dogma of *creatio ex nihilo* than by a *creatio ex profundis*.[37]

In the darkness of our own energies, in our depressions, both personal and collective, we have the opportunity to face into that depth. And we discern that it is not a blank emptiness. Rather than the nauseating thrust endlessly outward, the dark energy loops back inward as intensity. The deep is no cause for nihilism. It glows enough to disclose our entanglement in a planetary web, badly wounded, but nonetheless still replete with life. This pulsing life may energize a work that not only heals damage already done but evolves a new body as well. That body is each of ours. It can be lived as a soulless matter powered by captive energy or, alternatively, as the soulful flesh in which we know ourselves as members of one another's body.

Therefore we experience joy and sorrow in one another's joys and sorrows. By thus alluding to Paul's famous figure of the communal body, *Corpus Christi*, in his first letter to the Corinthians, I do not mean to send us to church. This misleadingly institutionalized metaphor has resonances that exceed the ecclesial. The "body of Christ" is raised in this metaphor from the single dead body of Jesus into an ongoing interactive manifold. It is its original Greek, that makes it of special interest to the present discussion. Paul repeatedly uses the language of energy, *energeia*, in this passage. The New Revised Standard Version of the Bible translates 1 Corinthians 12:4–6 thus: "Now there are varieties of gifts, but the same Spirit; and there are varieties of services, but the same Lord; and there are varieties of activities, but it is the same God who activates all of them in everyone." This is an improvement over the King James Version's "same God which worketh." But let us hew closer to the original version of this last verse: "And there are differences of *energizations* (*energema*), but God is the same One Who *energizes* all in all (*ho energon ta panta en pasin*)." Not a God who works on us from outside, who does our work for us, but who energizes our inter-activities. Not a God above pulling strings, but the energy of our intimate/infinite entanglement? This spiritual body, identified at once (and at the same time) as the church and the Christ, appears to be woven of interdependent "energizations," throbs of actualization, emotive, gifted, and alive.

Might the spirited community with its luminous energies of resurrection have provided, for a time, a workshop in the greater entanglement? Might its subtle body have hosted an interactive excess of corporeal life? What Ivone Gebara calls our "Sacred Body"—avoiding the dogmatic demands both of the body of God and of Christ—might now energize planetary hope, not Christian conversions. Such corporeal theology aims to practice a local mindfulness of bodies, of bodies within bodies, within the nonlocal universe that each of us uniquely embodies. Along with the polar bears. It is not all about us. It is about all of us. This seventh pulsation signifies a space in which the prior six may—"all in all"—be energized as desire, not dread. We take Blake's hint—we heed the body. We feel, amid our weirdness and our woes, the throb of eternal delight. We feel the energy we are.

# CHAPTER 2

# The Fire Each Time:
# Dark Energy and the Breath of Creation

*Mary-Jane Rubenstein*

## In the Beginning

The hot big bang hypothesis can be traced back to 1929, when Edwin Hubble discovered that the universe is expanding.[1] While Hubble observed this phenomenon directly, the possibility of an expanding universe had already been opened by Einstein's theory of general relativity, which he completed in 1915. According to general relativity, space and time are neither independent nor static substances; rather, they compose a dynamic "space-time" that can grow, shrink, bend, and warp in relation to matter and energy. Prior to Hubble's observational confirmation, then, Einstein's own theory suggested that the universe might either be expanding or contracting—that space-time itself could either be racing outward or retreating inward.

His own calculations notwithstanding, this possibility was notoriously difficult for Einstein to abide. He believed—partially out of adherence to the law of conservation of mass-energy and partially out of what one might call an ontotheological yearning—that the universe must be static. Something must be regulating the elasticity of space-time, or else gravity left to its own devices would cause the universe to collapse. Einstein therefore posited a counter-balancing negative pressure ($\Lambda$), which he called the "cosmological constant," to offer an equal and opposite push to gravity's pull. With gravity and lambda in perfect proportion, Einstein was able—theoretically, at least—to keep the universe from stretching out or caving in.[2]

Upon hearing the news of Hubble's discovery—that space-time was in fact expanding outward—Einstein immediately retracted his cosmological constant, calling it his "biggest blunder."[3] In the meantime, the work of a young Belgian physicist and priest named Georges Lemaître was vindicated. Two years before Hubble's discovery, Lemaître had posited an expanding universe based on Einstein's equations, provoking Einstein to tell him, "your math is correct, but your physics is abominable."[4] Encouraged by his colleagues' sudden interest in his work after 1929, Lemaître went on to suggest that if the universe is expanding now, then it must always have been expanding—from the beginning of things. The scientist-priest, soon to be made monsignor, went further: the whole universe, in all its multiplicity, must have exploded forth from *one* point, which he called the "Primeval Atom."[5]

The story goes like this: one day—although words fail us here, as there were no days—14 billion years ago, this Primeval Atom of infinite density issued forth a searing white light (and lo, there was light). The light radiated from an ultra-hot, ultra-dense plasma and formed "vast quantities of energy in the form of a dense particle soup."[6] This particle soup—a roiling, undifferentiated, dare one say tehomic gloop[7]—of quarks and gluons took 380,000 years to cool down sufficiently for matter to emerge. This cooling is attributable to a regime change that took place 75,000 years after the big bang, when the fires of radiation gave way to the gentler tug of gravity. Out of the plasma, gravity began to draw atoms and molecules to itself, coaxing galaxies, planets, and stars into being. Out and back: the universe flung into flaming infinity and then gradually attracted and assembled by the power of gravity. From Hubble's shocking discovery through the last few years of the twentieth century, this was the operative theory: the universe was born in a moment of insanely rapid and infernally hot expansion and was gradually given over to the demiurgic force of gravity, which has *slowed down the expansion of the universe ever since*:[8] a benevolent cosmic girdle.

And then the whole thing came ungirdled. In 1998 two independent research teams each discovered that the universe's expansion is not slowing down; it is speeding up.[9] Using type Ia supernovae to measure the brightness—and therefore the distance—of cosmic bodies, teams led by Saul Perlmutter and Brian Schmidt both found that faraway galaxies are not only moving farther away; they are moving farther away *faster* than they used to.[10] NASA's Wilkinson Microwave Anistrophy Probe (WMAP), launched in 2001 to gather light from the outer edge of the cosmic sphere, confirmed as much: the universe immediately after the big bang was 13.7 million light-years in radius. It is now 13.7 *billion* light-years in radius. At this rate the universe will double in size every 10 billion years.[11]

So Einstein was not entirely wrong. There *is* a counter-gravitational force, but it does not exist in a happy homeostasis with gravity. Rather, it won its cosmic tug-of-war against gravity 9 billion years after the big bang, which is to say 5 billion years ago. Since then it has been pushing space-time itself outward with increasing velocity. Of course, as Dan Hooper of the Fermi National Accelerator Laboratory reminds us, it is strange that anything should be expanding the space of the universe itself, because "by definition, there is no space outside of the Universe for it to expand into."[12] So something is pushing the world out into nothing at all. Shortly after its discovery in 1998, Michael Turner of the University of Chicago named this strange force "dark energy."[13]

## All Things Invisible

Particularly for the apophatically inclined, "dark energy" is a captivating name, and indeed, like all the names for the "divine darkness" of Pseudo-Dionysius, this one doesn't quite name what it names.[14] The "dark" of *dark energy* means two rather simple things: cosmologists can't see it, and they don't understand it. Hooper claims it is "the biggest mystery in all of physics," and Turner has called it "the most *profound* problem in all of science."[15] Like the depths of darkness perhaps more familiar to theologians, dark energy is unfathomable.

In addition to marking its inscrutability, calling this energy "dark" also connects it to the stubbornly elusive "dark matter" discovered in the mid-twentieth century. In the early 1950s a twenty-two-year-old graduate student named Vera Rubin realized that galaxies seem to be spinning too quickly to stay in orbit. They must, she suggested, be more massive than we think they are; there must be some invisible substance that weighs them down and prevents them from spinning into oblivion. A similar idea had been posited by the Swiss astronomer Fritz Zwicky in 1933, but Zwicky was notoriously crotchety; Rubin was young, and a woman, and the idea was odd. It was not until the 1970s that the mainstream scientific community realized that however odd, dark matter exists; moreover, it surrounds most galaxies in thick, invisible rings called *halos*. But unlike ordinary halos, these do not shine. The "dark matter" composing them neither absorbs nor releases light, nor does it interact with electricity or magnetism. It is hoped that some particles of dark matter might eventually be generated by the Large Hadron Collider in Geneva, but for the moment, scientists have never seen, felt, or heard dark matter, nor do they know what it is. And yet, to draw upon an old scholastic distinction, they know *that* it is.

As for dark energy, it has proven even more recalcitrant. Michael Turner has ventured the explanation that dark energy is kind of like dark matter, except "more energylike."[16] The ineptitude of this qualification seems deliberate; scientists simply do not know what the stuff is. In the words of David Schlegel of the Lawrence Berkeley National Laboratory, "the term doesn't mean anything. It might not be dark. It might not be energy. The whole name is a placeholder for the description that there's something funny that was discovered [twelve] years ago now that we don't understand."[17]

As is evident from Schlegel's mild peevishness, "not understanding" dark energy is particularly troubling to physicists. This is primarily because, whatever it "is," dark energy seems to compose the majority of "is-ness" itself. According to NASA, of all the universe's mass-energy, approximately 70 percent is dark energy, about 25 percent is dark matter, and less than 5 percent is visible matter.[18] That is to say that all we can see—tables, puppies, pencils, planets, and stars—everything that seems to *be*, is on balance almost nothing.[19] Lawrence Krauss of Case Western Reserve University has therefore called the discovery of dark energy "the ultimate Copernican Revolution." We will recall that Copernicus unsettled Christendom's fantasy that the earth was the center of the solar system. Subsequent centuries of astronomy have shown that our solar system is not the center of the galaxy, nor is our galaxy the center of the universe. Far from being the center of anything at all, Krauss says, "we're just a bit of pollution," nowhere in particular.[20]

Now, any student of European history knows what a difficult time the Church has had with Copernican Revolutions—from Galileo to Darwin to Freud (who believed his Copernican Revolution to be the last and greatest). Dogma has been on the defensive since the birth of modern science. What is surprising, however, is the extent to which dark energy's Copernican Revolution has unsettled cosmologists themselves. There are a number of reasons for this, most of which amount to dark energy's multifarious obstruction of a Theory of Everything—that is, an account of the four forces of the universe that would harmonize quantum theory with general relativity.[21] Infamously, when quantum mechanics calculates the value of dark energy, it produces a number $10^{120}$ times larger than the amount of dark energy there actually seems to be (for comparison, there are only about $10^{80}$ atoms in the entire visible universe).[22] This has been a massive embarrassment to the cosmological community. "It's been so hard," Perlmutter has confessed, "that we're even willing to consider listening to string theorists. They're at least providing a language in which you can talk about [quantum mechanics and general relativity] at the same time."[23] What is troubling about string theory is that it can end up positing anywhere between $10^{500}$ and $10^{1000}$ *types* of potential universes. "Listening to string theorists," in other words, opens the possibility that the universe is not one. It might well be a multiverse. And as Mark Livio of the Space Telescope Science Institute (STSI) has admitted, the idea of "a zillion universes" "raises the blood pressure of many physicists seriously."[24]

Dark energy's ability to provoke such anxiety is rooted, it seems, in the possibility that the incomprehensible part of the universe might not only dwarf but also ultimately swallow what is a comprehensible whole. At this point in space-time, dark energy constitutes a bit more than 70 percent of the universe. If the dominant hypothesis is correct, dark energy will cause the universe to stretch outward until, a few trillion years from now, "every individual galaxy, star, and planet will be ripped apart" or gradually dissolved and nothing but dark energy will remain.[25] So the invisible might overrun the visible; darkness might, in fact, overcome the light that shines in it. Yet it is for this very reason that Michael Turner insists, "we can't understand the universe until we discover what dark energy is."[26] Another familiar theologeme: we cannot understand it, but we have to understand it. And so it is that Lawrence Krauss told a group of physicists and astronomers at a recent meeting at STSI: "in spite of the fact that you are liable to spend the rest of your lives measuring stuff that won't tell us what we want to know, you should keep doing it."[27]

## *Energy and Unknowing*

When faced with the problem of knowing something that "won't tell us what we want to know," the Eastern Orthodox tradition has recourse to a term with which the Western tradition is largely unfamiliar: the *energy* of God. This line of thinking was most clearly set forth by Gregory Palamas in the thirteenth century and revived by Vladimir Lossky in the twentieth, but it can be found in the fourth-century Cappadocians as well: God is unknowable in essence (*ousia*), but knowable in energy (*energeia*).[28] The divine energies are said to be the Pseudo-Dionysiusian *proödoi*—the emanations of God—or as he often puts it, "the ray" issuing forth from the divine darkness. The unknowable divinity makes itself known in the energies yet remains hidden in essence. The essence is dark, but the energy is light.[29]

As Lossky explains it, this distinction between essence and energy emerged as a corollary to the doctrine of *theosis*, or deification. According to Orthodox teaching, the soul can participate through grace in the very life of God. Now, considering God's radical transcendence of creation, this teaching seems puzzling. How can the soul gain access to the life of a wholly inaccessible God? How can it take *part* in a God whose simplicity renders God indivisible into parts? As Lossky explains it, humans cannot, in fact, participate in the essence of God, for that would effectively divide God among the participants. Nor can humans participate in the persons of the Trinity (*hypostaseis*), for the hypostatic union belongs to the Son alone. And yet 2 Peter assures us that we shall become "partakers of the divine nature."[30] Lossky concludes, "We are therefore compelled to recognize in God an ineffable distinction . . . between the essence of God, or His nature, properly so-called, which is inaccessible, unknowable, and incommunicable; and the energies or divine operations, forces proper to and inseparable from God's essence, in which He goes forth from Himself, manifests, communicates, and gives Himself."[31] God therefore exists in two modes: *en ousia* (in essence), God is known only to Godself, while *en energeia* (as energy), God communicates Godself outside Godself.

At this juncture one might be tempted to map the distinction between essence and energy onto the distinction between the "immanent" Trinity (which expresses God eternally within the Godhead) and the "economic" Trinity (which expresses God temporally out to creation). But this would not be quite right: the energies are eternal and therefore independent of creation. And the energies must be eternal because, as Lossky maintains—although it is unclear how he knows—God *would* manifest Godself beyond Godself even if there were no creation.[32] Yet here one might ask: what would "beyond Godself" mean if there were no creation? In an effort to circumvent this problem, Lossky has recourse to Palamas's rather hasty distinction between energies that *are* tied to time (creation, redemption, providence, etc.) and those that are eternal (wisdom, power, goodness, etc.).[33] But this dichotomy risks compromising the unity of the *energeiai* and introducing an unorthodox shift from potentiality to actuality within the Godhead, which from eternity is held to be fully actualized. In short, the energies occupy a rather fuzzy ontological position. Lossky claims this fuzziness for orthodoxy by dubbing it "antinomic"; just as Christ is both fully human and fully divine, the energies are both free from creation and bound up with it. On the one hand, Lossky explains, "they belong to theology [the immanent trinity], as eternal and inseparable forces of the Trinity existing independently of the creative act; on the other, they also belong to the domain of 'economy,' for it is in His energies that God manifests Himself to creatures."[34]

Leaving aside for a moment their categorical instability, the "energies" designate all the divine attributes: Wisdom, Life, Power, Justice, Love, Being, Goodness—all of the terms Pseudo-Dionysius enumerates in his treatise on the *Divine Names*—including, one imagines, "worm," "drunk with a hangover," and far more problematically, "essence" (I shall return to this momentarily). All of the names of God are energies—even "God," according to Gregory of Nyssa, insofar as it names the faculty of governance. "We name Him from each of them," Palamas says of the *energeiai*, "although it is clear that He transcends all of them."[35] And indeed, this is in keeping with Pseudo-Dionysius's conviction that as Creator, God can and must be named by all creation, but that as Creator, God also exceeds creation.[36] Having named God with every name, Dionysius therefore entreats us to *unsay* them, denying the names from the lowest (*worm, drunk, rock*) to the highest (*Father, Son, Holy Spirit*) until, unknowing, we "may see above being that darkness concealed from all the light among beings."[37]

For Dionysius, the divine names, or energies, both reveal God to creation and conceal God from it. This is the reason they must both be affirmed and denied. So in ascending past the "ray" of the names to the Darkness of God, do we thereby move beyond energy to essence? In this passage from the *Mystical Theology* at least, Dionysius seems to indicate as much; once we have denied the highest of revealed names, we will contemplate the concealed God. Yet Palamas and Lossky expressly deny this possibility, asserting that the deified soul in Dionysius finds its end *not* in the Darkness of God but rather in the uninterrupted light of energy.[38] In short, Palamas and Lossky reverse the apophatic privilege of darkness, relegating it to a fleeting moment on the way to light. Furthermore, they lose sight of Dionysius's insistence that the divine *exceeds* both darkness and light, and that the soul is only united to God once it is carried beyond this opposition. This puzzling light-supremacy can perhaps be most clearly seen in Palamas's and Lossky's conviction that, just as the dark revelation of the *Mystical Theology* purportedly gives way to the uninterrupted light of the

*Divine Names*, the darkness of Sinai has been overcome by the Light of the Transfiguration.[39]

While it is beyond the scope of this exploration to perform a systematic deconstruction of the distinction between essence and energy,[40] we have thus stumbled upon a number of places in which it trembles. In order to align essence with darkness and energy with light, Lossky and Palamas must reconfigure the apophatic journey so that it ends, not in darkness, nor in neither-light-nor-darkness, but in uninterrupted light. Second, their designation of Mts. Sinai and Tabor as indexes of this progression requires not only a healthy dose of Christian supercessionism but also a rather flat-footed reading of the Transfiguration (after all, the voice of God comes not from unending brightness but from a "bright *cloud*" that "overshadows" them).[41] Third, to call upon Pseudo-Dionysius in support of the *ousia/energeia* distinction is questionable at best, insofar as he names *"ousia"* among the divine names that Palamas and Lossky equate with the *energeiai*.[42] Surely, if essence *is* an energy, it is not entirely distinct from energy. And finally, insofar as some of Dionysius's divine names are tied to creation (*creator, redeemer, sustainer*, etc.), Palamas is forced to admit that some energies are not eternal, even though most of them are. And again, "most of them are," or must be, because of the speculation we saw earlier in Lossky: God *would* have revealed Godself outside Godself—even independently of an "outside" to which God might communicate—because it is in God's nature to exceed Godself. What all of these faltering moments seem to indicate is that the divine essence *is some sort of energy*. Certainly such a divine essential energy would exceed the human capacity to discern it, but insofar as any discernment or participation is possible, it would be by virtue of an outpouring of this energetic essence. This is nothing more than what Plotinus ventured in synthesizing the Platonic One with the Aristotelian Intellect: energy flows forth *from* God (*energeia ek tês ousias*) out of the energy *of* God (*energeia tês ousias*).[43] Like fire, God *gives* energy because God *is* energy.

Along this reading, the divine energy would be the very stuff of God, pouring outward and flowing back. Just as the divine names propel both the cataphatic descent and the apophatic return, divine energy works in equal and opposite directions to push creation out and pull it back—like a carefully calibrated cosmological constant. Dionysius called these two functions *proödos* and *êpistrophê*; Meister Eckhart called them *effluxus* and *refluxus*. At the same time that it fires forth into creation, the divine energy gathers creation back to itself, until creation itself is, or as Eckhart would put it, until everything is fire.[44]

## *In the End*

The discovery of dark energy therefore poses the same problem for theology as it does for physics: it threatens to throw the cosmic scales off balance. As Jacques Derrida has pointed out on numerous occasions, even "the most negative of negative theologies" knows where it's going and goes where it comes from.[45] Even the most tortuous path of Dionysian denials is still a *path*, starting and ending with the alpha and omega who processes outward to draw all things home. But dark energy seems to obliterate this path, leaving us with an excess of *proödos* over *êpistrophê*: with an outpouring that does not, or will not, or cannot, gather itself back.

This dominant theory is called "the inflationary hypothesis," first posited by Alan Guth in 1979.[46] A patchwork of twentieth-century discoveries, the theory of inflation integrates the big bang hypothesis from the 1930s with the dark matter discovered in the middle of the century, the "inflationary energy" championed by Guth in the 1980s, and the dark energy discovered in the 1990s.[47] The new, improved story goes like this: 14 billion years ago, space and time fired out from a tiny nugget of imponderably high density. After a millionth of a second, this nugget expanded into a flat and smooth cosmic terrain, 13.7 million light-years across.[48] This sudden transition is attributable to a mysterious force of nearly infinite power called "inflationary energy," which appeared for a flash and then disappeared forever.[49] Three cosmic stages have ensued: the first dominated by radiation, the second dominated by matter, and the third dominated by dark energy (36). During the radiation phase, matter and antimatter erupted and canceled each other out, with matter gaining a slight advantage over its opposite. Quarks and gluons bound to form protons and neutrons, and the whole world was a hot-as-hell cosmic soup. As gravity gradually overcame radiation, the primordial plasma cooled and atoms emerged. Gravity attracted matter to cores scattered throughout the universe, and stars, planets, and galaxies began to form. Then, 9 billion years after the big bang, dark energy's push began to exceed gravity's pull, and the universe's expansion sped up. With the demise of gravity, no more cosmic bodies will be created. From now on, the universe will continue to accelerate until galaxies, stars, and matter itself are unbound, either ripped apart or dissolved into a void of dark energy—an end that some physicists have dubbed "the Big Whimper."[50]

It seems to me that this whimpering end of inflation reveals the secret ontotheological yearnings of all of us; it's no wonder it gives scientists high blood pressure. Nor is it any wonder the numbers don't work out. At the moment that dark energy overran gravity, the whole cosmos seems to have been let loose with no ticket home. What we have here seems a terrifying excess of the incalculable over the calculable, of procession over return, of Abraham over Ulysses, of Kierkegaard over Hegel. To the chagrin of seekers of orthodoxy and a Grand Unified Theory alike, it seems that God might indeed play dice. In fact, God seems to throw the dice outward, beyond even God's own reach.

But even in a Kierkegaardian universe, there is always an "or."

## *Or in the End*

In the five years since the WMAP results emerged, an alternative hypothesis has been taking shape through the work of two theoretical physicists, each of whom was instrumental in setting forth and developing the inflationary hypothesis they now jointly contest (6).[51] Paul Steinhardt and Neil Turok's alternative cosmology was born out of their three-pronged dissatisfaction with the dominant model. First, this model seems to them a "patchwork" of different elements, periodically adjusted to accommodate a newly discovered force, but with "no overarching principle" (66). Second, in order to explain how the universe gets so big and flat so quickly, it has to posit "inflationary energy—an impossibly explosive force that lasts for an instant and then disappears forever." And third, while the inflationary hypothesis can account for the events after the big bang, "the big bang itself is not explained. It is simply imagined that space and time emerged somehow" (6). In other words, the inflationary hypothesis posits a big bang *ex nihilo*, and as Steinhardt and Turok put it, "there are no rigorous physical principles that dictate how to go from 'nothing' to 'something'" (226).

We will recall that the father of the big bang theory, in addition to being an internationally renowned physicist, was a Roman Catholic priest. It is said that Lemaître took pains to keep his science independent of his theology, and perhaps for this reason (or perhaps heeding the warning of St. Augustine's colleague[52]), he never ventured an opinion as to whether there was anything "before" the big bang. Nor was there any scientific consensus on the matter in 1951 when Pope Pius XII nonetheless declared it to be in perfect line with Catholic doctrine. "It would seem," Pius told the Pontifical Academy of Science, "that present day science, with one sweep back across the centuries, has succeeded in bearing witness to the august instance of the primordial Fiat Lux, when along with matter, there burst forth *from nothing* a sea of light and radiation, and the elements split and churned and formed into millions of galaxies."[53] In the decades since then,

Steinhardt and Turok claim it has become commonplace among cosmologists simply to *assume* the big bang produced space-time out of nothing,[54] but apparently there is no more physical evidence of such a thing than there is biblical evidence of it.[55]

Over against the inflationary ex nihilo, Steinhardt and Turok posit their alternative: a cyclical universe, in which the big bang is not the beginning of time or space. Rather, it is a brief flash of cosmic renewal that takes place every trillion years or so. "In each cycle," they explain, "a big bang creates hot matter and radiation, which expand and cool to form the galaxies and stars observed today. Then the expansion speeds up, causing the matter to become so spread out that space itself approaches a nearly perfect vacuum. Finally, after a trillion years or so, a new big bang occurs and the cycle begins anew" (6). This alternative model addresses all three alleged inadequacies of the inflationary hypothesis.[56] The cyclical world is eternal; hence there is no need to jump from nothing to something. The universe remains relatively large, flat, and smooth throughout the cycles, so there is no need to posit "inflationary energy." And unlike the "patchwork" composing the inflationary model, the cyclical model does have an "overarching principle": dark energy guides the whole process.

Steinhardt and Turok's story more or less follows the trail of the accepted hypothesis, except without the inflationary energy. First there is a big bang, then radiation, then matter, then dark energy, but at this point, when the dominant model posits the gradual unraveling of the world, the cyclical model changes course. Dark energy does not continue to speed the cosmos outward forever, either ripping it apart or dissolving it into a void. Rather, after a trillion years dark energy begins to decay.[57] Its outward force starts to reverse, propelling the cosmos into "a phase of very gentle contraction" (63). This contraction draws the universe into a "big crunch," at which point another big bang bangs and the universe is thrown outward again. Perhaps the most significant ontological distinction between these two models, then, is that while the inflationary hypothesis maintains that creation ended with the onslaught of dark energy and will eventually be entirely unmade, the cyclical hypothesis promises (in rather hymnic syntax) that after each bang, "created anew will be galaxies, stars, and planets like Earth on which intelligent forms of life may develop" (65).

According to the cyclical model, the whole world seems something of a phoenix, periodically undone and remade in fire. And in fact the first name Steinhardt and Turok gave to their hypothesis was *ekpyrosis*: out of fire.[58] They borrowed the term from Stoic cosmology; as Cicero's Balbus explains in *On the Nature of the Gods*, the early Stoics taught that "there will ultimately occur a conflagration of the whole world. . . . nothing will remain but fire, by which, *as a living being and a god*, once again a new world may be created, and the ordered universe restored as before."[59] We will note that in the Stoic cosmos, the god who creates, orders, and consumes has no essence in excess of its energy: the god *is* this fire itself. Similarly, for Steinhardt and Turok, the role is played by dark energy. As the "overarching principle" of the cyclical model, dark energy "regulate[s] the cycling" by causing cosmic expansion, stabilizing the universe, and absorbing systematic shock (68, 241). What this means is that all things visible, "stars, galaxies, and the larger-scale structures observed in the universe today *owe their existence* to the period of dark energy domination in the previous cycle. And the dark energy *dominating* the universe today is *preparing* similar conditions for the cycle to come" (67; emphases added). So rather than unmaking the world once it overtakes gravity, the dark energy of this model is dominant the whole time and creates, by virtue of this perpetual dominance, and executes the whole affair according to a plan.

With the exception of this sudden anthropomorphism, the cyclical model recapitulates not only Stoic *ekpyrosis* but also Hindu cosmology, theories of Buddha worlds, not to mention Nietzsche's eternal return.[60] The major difference lies in the last major component of the cyclical model, which is that our universe is not the only universe. There is one more. We will recall that Saul Perlmutter, leader of one of the teams that discovered dark energy, said that cosmologists were "even willing to listen to string theorists" in order to reconcile quantum mechanics and general relativity. And indeed the cyclical model draws its inspiration from the mysteriously named "M-theory," which Edward Witten proposed in 1995 as an integration of all existing string theories.[61] Steinhardt and Turok's interpretation of M-theory posits not $10^{500}$ universes, but two.[62] According to this hypothesis, the whole world—from our hands, to the stars, to the remotest galaxies—exists on a flat, extended membrane, or "braneworld." Across a tiny gap along an imperceptible dimension there is another braneworld, filled with all sorts of matter and energy as well. In fact, the reason "dark matter" exerts such force on our cosmos without being visible *could* be that it is lying on this other world (140). The brane next door hovers a hairbreadth away—"perhaps 10–30 cm across," but we cannot see, touch, or hear it. "We are stuck like flies on flypaper," Steinhardt and Turok explain, "and can never reach across the gap to the 'hidden' world, which contains a second set of particles and forces with different properties from those in our braneworld" (139). Nearer than hands and feet, it remains totally inaccessible. Nothing can reach across this gap—except dark energy.[63] Dark energy shuttles along the fourth dimension, pushing the two branes apart and then drawing them closer together.

What this means for the cyclical universe is that the big bang does not produce a world that is 13.7 million light-years in radius out of a single point. Rather, it marks the collision of the two braneworlds (143). This collision produces the familiar primordial soup and radiation *on each braneworld*, matter eventually gathers and forms on each braneworld, and then dark energy prompts the expansion of space-time along each braneworld, while also increasing the distance between them. As the force of dark energy reverses itself, it causes the branes to contract slightly and draw closer to each other, ultimately colliding in a big crunch: "The collision between two branes would produce a searing white light, signaling the beginning of a new cycle of cosmic evolution" (193). And another big bang bangs, and the world is born again out of red-hot plasma.

## Flints

In this contemporary debate between the inflationary and cyclical hypotheses, we seem to have two models of creation that do not quite fit any of the old theological standbys.[64] To be sure, each of them contains familiar elements, but in combination with unexpected partners. The inflationary hypothesis gives us a world created ex nihilo, in good orthodox fashion. At the same time, it presents a future without redemption or renewal, abandoned to a growing void that will ultimately consume it. The cyclical model, on the other hand, begins from the heretical conviction that nothing comes from nothing, so the world must be eternal. But the force that oversees this eternity "prepares" cycles in regular intervals, conserves the world's mass-energy, and all in all, issues a fairly ontotheological guarantee of procession and return. So the inflationary Elohim breathes out, and out, once and for all. The cyclical Elohim breathes out, and in, and the cosmos is reborn.

The Eastern churches teach us that one participates in the very life of God, dwelling in the Inaccessible, by means of the divine energies. Yet how exactly does one attain such participation? We will recall that the *energeiai* are wisdom, power, goodness, beauty, truth—all the manifestations of the hidden God. As Aristotle Papanikolaou explains, the energies are "the bridge of the unfathomable gap between the uncreated God and God's creation."[65] To share in these energies would therefore mean to receive them as they cross this unfathomable gap and to reflect them back across and out into the world. The crucial passage here is from I Corinthians:

> Now there are diversities of gifts, but the same Spirit. And there are differences of administrations, but the same Lord. And there are diversities of operations [*energêmatôn*], but it is the same God which worketh [*ho energôn*] all in all. But the manifestation of the Spirit is given to every man to profit withal. For to one is given by the Spirit the word of wisdom; to another the word of knowledge by the same Spirit; To another faith by the same Spirit; to another the gifts of healing by the same Spirit; To another the working of miracles; to another prophecy; to another discerning of spirits; to another divers kinds of tongues; to another the interpretation of tongues. But all these worketh [*energei*] that one and the selfsame Spirit, dividing to every man severally as he will.[66]

Each of the spiritual gifts is an *energema*: the Holy Spirit gives them so that creation might mirror the divine energies. We might therefore think of the divine energies as a call and the charisms or virtues as a response: God acts, and by virtue of our reception of this action, we are enabled to act in like fashion, imitating God's love, healing, wisdom, power of reconciliation, and *creation*—to such an extent that we share in the divine life itself.[67]

This is almost right, except, as David Bradshaw has argued, the workings of the Spirit are a bit more complicated than action and reaction. In a number of highly systematic studies, Bradshaw has shown that every one of Paul's twenty-six uses of *energeia* and *energein* describe the actions of God, Satan, or demons—never of humans.[68] Or never of humans alone. One example Bradshaw gives is Colossians 1:29, "where Paul refers to himself as 'striving according to Christ's working [or energy, *energeia*], which is being made effective [or energized, *energoumenên*] in me.'" "On the one hand," Bradshaw explains, "the divine energy is at work within Paul, transforming him, so that from this standpoint he is the object of God's activity; on the other it finds expression in Paul's own activity, so that Paul's free agency and that of God *coincide*."[69] So it is not quite the case that God works and then humans work back. Rather, God works in me in such a way that my working *is* God's working: the deified creation does the healing and reconciling and creating *along with God*—to such an extent that the actors, in a sense, merge. Bradshaw calls this strand of Paul's theology "synergistic": "The energies are precisely the realm of *reciprocity*, that in which God shares Himself [sic] with creatures and summons them to offer themselves to Him."[70] Thus offered, humanity becomes the venue of the divine energy; as Eckhart might explain it, "God is his own place to work in."[71] Insofar as it relates to creation, then, the divine *energeia* is *synergeia*.

But here we tread on dangerous ground. Here we risk making the divine operation somehow dependent upon creation's participation in it. It is for this reason that the Orthodox—Bradshaw included—install a double-glazed barrier between the theotic creation and the *theos* itself. The first layer to this barrier is the eternity of the energies: God is said to manifest Godself, internally *and externally*, independently of creation. Yet we have seen two major cracks in this pane: first, not *all* of the energies can be said to be eternal ("creation" being chief among them), and second, without creation there would be no place to which God might manifest Godself. Hence the second layer exists to contain the leaks of the first: the divine energies, it is said, are not the divine essence. So even if the *energies* get muddled up in the mess of the world, there is still a reservoir of divine ontology that remains uncorrupted, untainted—totally removed from creation. Again, however, we have seen this distinction collapse as well, in short because of the pesky conviction to which Lossky himself gives voice: God cannot not express Godself. Which is to say God's essence, in all its unknowable darkness, *is* energy.

What happens, then, if the divine essence is energy and the divine energy is synergy? (And here I can almost feel the flames at the base of my heretic's stake.) This, according to Rowan Williams, is the greatest danger the *energeiai* (which he uses interchangeably with the Dionysian *dynameis*) present: they seem to evacuate the contingency of the world. As Williams demonstrates in his own dismantling of the *ousia/energeia* distinction, "it is hard to avoid the conclusion that the world must be eternal, insofar as the *dynameis* are eternally engaged, by their very nature, in *communicating* the divine perfections to some second term or order of being. God and the world appear to be bound up in a kind of organic unity—a foreshadowing of Whitehead or Hartshorne."[72] Williams tries to avoid this unsavory prospect by means of Aquinas's equation of essence and energy in the Godhead, establishing God as pure actuality and creation as its contingent recipient.[73] But it seems to me the damage has been done: if the energies cannot be dislodged safely from the divine essence *or* from creation, then they do seem to bind the whole lot into an eternal, synergetic dance. This is not to say that God is creation or that creation is God, but by virtue of the eternal energy that "bridges" them, it is possible that both become, and unbecome, in mysterious relation to each other.

Across an imperceptible divide, a strange fire flings this world away from that one and then slowly joins them back. So our universe is haunted: what we know is there won't quite show itself. But a dark force races between worlds, drawing us into the rhythm of creation; breathing out and breathing in. Thrown outward in a sea of light, we are gathered back by the darkness, from fire to fire. "And the dust returns to the earth as it was, and the spirit returns to God who gave it," and "the quarks and gluons of which we are all made join the flood of new quarks and gluons created at the bang, and the cycle of the cosmos is renewed."[74] World without end. Perhaps.

# CHAPTER 3

# Solar Energy:
# Theophany and the Theopoetics
# of Light in Gregory of Nyssa

*T. Wilson Dickinson*

In the seemingly material problems that surround the word energy, I cannot help but hear the metaphysical echo of the Greek term *energeia*.[1] In the faint resonance I hear between these two terms there seems to emerge a rather unlikely pairing: that of the looming ecological crisis and that of the ancient disciplines of philosophy and theology. Though at first this pairing might seem forced, in this essay I argue that the possible melodious performance of these two unlikely partners might very well provide a way forward. While many will propose that solar energy might be the most promising resource to avoid looming ecological catastrophes, I explore the transformative possibility that a different kind of solar energy—that of the theopoetics of light—could have.

In a context in which the shortage of energy threatens to destabilize any semblance of geopolitical order but the previous misuse of energy also threatens to destroy the ecological stability of the earth, the lengthy consideration of this ancient and theological resonance within the term energy may sound like the idle chatter one might expect from theologians and philosophers. The pragmatically oriented person might charge that by giving in to these theological curiosities in the face of real problems that threaten the conditions of possibility for our lives, I have run away from the physical world so as to find everlasting stability in the metaphysical. Though I imagine I cannot entirely escape the risk that my inquiry will ultimately be found guilty of such accusations, I am not so sure that the metaphysical can so easily be separated from the physical. I am sympathetic to the Heideggerian claim that the processes of modernization are perhaps the outgrowth of a metaphysical tradition that has become radicalized.[2] The echoes between the English energy (with all of its technological, journalistic, and informational overtones) and the Greek *energeia* (with its metaphysical and theological pitch) might be more significant than a shared linguistic resonance. Perhaps the crises of energy are as much the result of "material" and "practical" situations as they are the consequence of a common outlook. The crises currently surrounding energy might not only threaten the preservation of our lifestyles and ways of thinking, but perhaps these problems are derived from them as well.

In light of this connection, it seems that the emerging energy crisis and the corresponding ecological crisis need to be addressed by a theological and philosophical shift in thinking. In invoking the dual sense of energy, I am not attempting to argue for a causal relationship between the two, nor will I attempt to show the underlying logic of our contemporary situation through its ancient roots. Instead, I am wondering how the cacophonous resonance of these two terms might reveal both a shared problem and the possibility for a transformation or an alternative performance of this score. The hope is that this can be done without reducing this "practical" problem to a "theoretical" issue. I am wary of imagining that all that is required in the face of destruction, scarcity, and suffering is righting the course of the metaphysical ship. To do so would seem to imagine that theology and philosophy are the intellectual captains of the ship, or, to shift into a more contemporary technological metaphor, that the practical applications of the world are underwritten by a kind of theoretical "binary code."

While I would want to argue for a privileged position for theology and philosophy, I would want to do so by emphasizing the role they play as practices, as performances. Though my concerns here are deeply theoretical, they are understood as such not because they are intellectual or metaphysical rather than sensible or physical, but because of the resonance that the ancient Greek word for theory, *theoria*, has with vision. In its older usage *theoria* was largely understood in relation to the practice of witnessing and testifying to foreign and religious events and spectacles.[3] Viewed in this manner, theoretical discourse exceeds intellectual abstraction as it is part of an active embodied practice. What is at stake here is a matter of optics, of how one looks at the world, but it is an optics that cannot be situated as being purely on the side of the intelligible or the sensible. The resonance of these two terms, of the practical energy and the theoretical *energeia*, does not resolve itself with the complete priority of one over the other but instead seems to highlight the problematic configuration that would divide and order nature or reality by and into these two realms.

In the play between these two terms I propose that one can hear an alternative approach between a simply physical or material solution to the problem of energy, or a metaphysical escape into *energeia*. While the problems that surround the contemporary specter of "energy" will require scientific and engineered efforts, all such gestures will remain futile if the world is still viewed as simply being a collection of controllable objects and as the mere instrument of human desire. So long as our thinking and acting remains limited to an outlook of scientific determinism, a lifestyle of conspicuous and needless consumerism, and an ethics of atomized individualism, any practical efforts we might make to curb looming catastrophes seem to be doomed to failure.

Perhaps the problems that surround the shortage and overuse of energy, and the threats that accompany them, demand more than "practical" and "material" solutions. Perhaps what is needed is a transformation of the way many people have come to view nature and creation. Martin Heidegger laments the mode of thinking whereby "knowing subjects" have been so severely separated from their world that they view it as simply being the raw material that stands in reserve for their technological efforts. In this configuration the very capacity of our vision has become limited as the "earth now reveals itself as a coal mining district, the soil as a mineral deposit."[4] Pierre Hadot, citing Francis Bacon (among others), identifies this relationship with "nature" with the figure of Prometheus. Like the Greek son of a titan who steals fire from nature for humans, modern scientific thought, with "boundless curiosity, the will to power, and the search for utility," does not wait for nature to reveal herself, but takes her riches by its own means, and in so doing empowers the human capacity for knowledge.[5] In the imagery of early scientific texts, the Promethean outlook emphasizes the forceful and violent human capacity to tear the veil off of nature. Bacon holds that nature "unveils her secrets only under the torture of experimentation."[6]

In contrast to this outlook, Hadot outlines a train of thought and thinkers that might be represented by the figure Orpheus, whose optics is necessarily indirect, as he "penetrates the secrets of nature . . . through melody, rhythm, and harmony."[7] Whereas the former reduces its surrounding to being a collection of objects through the work of calculations and mechanized apparatuses, the latter emphasizes the creative aspects of all activities. The Orphic attitude sees its surroundings in terms of mystery and reverence rather than being oriented toward knowledge and possession. Johann Wolfgang von Goethe, whom Hadot identifies with this later disposition, frames the human capacity for knowing in creative, poetic, and artistic terms. Goethe emphasizes the centrality of art, which, unlike science, "does not discover hidden laws, equations, or structures behind phenomena; on the contrary, it learns to see phenomena, or appearance, what is in broad daylight, what is under our very noses, and which we do not know how to see. It teaches us that what is most mysterious, what is most secret, is precisely that which is in broad daylight, or the visible."[8] In outlining this distinction, I am not trying to exclude or villainize the so-called Promethean outlook. I am merely pointing out that it seems as though the same Promethean orientation that has caused many of the problems that are now invoked by the term *energy* is being deployed to overcome them. I propose that by learning from this other manner of looking at the world, through a reverent and poetic optics, we may be able to re-approach and transform our instrumental efforts.[9]

This essay is one of many expeditions in the search for "alternative energy." Specifically, it seeks to plumb the depths of the ancient wells of the writings of the fourth-century Christian theologian and philosopher Gregory of Nyssa. Though the term *energeia* is of central importance here, it is not confined to its traditional speculative and metaphysical parameters. Instead, *energeia* is considered as an opening for a theopoetic vision of creation. This consideration of *energeia* unhinges the binaries of the sensible and the intelligible and theory and practice that often govern the interpretation of Gregory's writings. In place of an emphasis upon a rigid and representational logic, in this essay I follow the pedagogical itinerary and poetic valence of his work, as his intelligible vocabulary is penetrated by his invocation of sensible images of light, and his theoretical vision is intertwined with a performative imperative. This shift in thinking, therefore, is not merely oriented toward a theoretical or academic object (the writings of Gregory of Nyssa) but also toward the transformative capacities that the "alternative energy" of theology and philosophy have for how one looks at and acts in one's world. As such, while this essay is directed toward a modern academic audience that seeks to understand Gregory, it is also an exercise of academic philosophy and theology in its ancient sense—seeking the formation of virtues above the transmission of information.

## Transformative Readings of Gregory

The "metaphysical" reading of Gregory's work—which maintains a number of dualisms and constitutes a system of distinctions—is not without grounding in his writing. For example, at certain points Gregory seems to echo the ontological chain that Jacques Derrida has described as governing the rationality of the division between the sign and the signified and the metaphysical binaries that accompany them. Derrida cites the opening passage of Aristotle's *De Interpretatione* as a prime example of this mode of thinking: "Now spoken sounds are symbols of affection in the soul, and written marks are symbols of spoken sounds."[10] In this passage, writing is shown to be derivative of speech, and speech to be derivate of the more direct and present thought. Therefore, the immediacy of the intelligible and the theoretical is considered to be superior and more original than the derivative domains of the sensible and the practical. Making one side of the binary primary or superior in this static fashion can serve to make the other both secondary and devalued (a particularly troubling tendency when one considers that the implication of devaluing the sensible is a devaluing of creation).

Gregory seems to outline his own ontological chain by "reasoning backwards" when he writes:[11]

> If we cannot first explain what is being said about God before we think it, and if we think by means of what we learn from his actions, and if before the act there exists the potency, and the potency depends on the divine will, and the will resides in the authority of the divine Nature— does that not make it clear to us that it is a matter of applying to the realities the terms we use to indicate what happens, and the words are a kind of shadow of the realities, matching the movements of things which exist?[12]

If one were to trace this backward movement forward, then one would see that the priority of God's being (*ousia*) flows from his nature (*physis*), through his will and power (*dunamis*) into his actions (*energeia*), which are perceived by thought, and only the latter may be derivatively represented by language. So long as this chain is maintained, *energeia* merely plays an intermediary role within a larger metaphysical frame, in which God's activities are the appearances of a deeper reality that is both unified and stable.

This particular formulation, however, is given in Gregory's lengthy polemic against the overly rigid and representational metaphysics of Eunomius. In this debate Gregory argues for the radical contingency of language and the insurmountable limitations of human knowing (*epinoia*). He argues at length that God's being and nature are beyond the human (and angelic) capacity for knowing. Therefore, the terms and even their intelligible referents on the higher part of the ontological chain do not have any positive content but merely possess a negative force. Even if one were to read this passage in a rather straightforward manner, it would be difficult to interpret the given chain of terms in too rigid a sense, as they do not positively designate something that is known.

So long as one presumes that Gregory's primary task is to outline a metaphysical system, his declared limitation of thought might merely be a mystification, a duplicitous bait and switch. If one places Gregory's texts in their "pedagogical" context, however, they seem to take on a different character. For example, the lengthy quote above that outlines the ontological unfolding of God is not simply declared. It is posed in the form of a question. Therefore, it could be read as a dialogical challenge and not merely as a representational statement.[13] In addition to having a dialogical character, this text is also part of a larger dialogue (however polemical) in which he is replying to the terms of a debate between two other people (his brother, Basil of Caesarea, and Eunomius). As such, many of the terms and formulations largely stem from the frames and parameters of his opponent in this debate.[14] The formulations of Gregory's text, then, can be understood not merely as assertions but also as replies to someone else's terms that often challenge the reader to think through their implications rather than merely to accept their validity blindly.

Though the modern tendency to understand historical theology in terms of doctrinal formulation has largely limited Gregory's contribution to the "development of Trinitarian doctrine," such interpretive frames need not be assumed.[15] In fact, it seems that Gregory himself outlined a very different itinerary. The organization of Gregory's work into theological and spiritual domains that so often accompanies contemporary doctrinally oriented readings seems to miss much of the performative and philosophical thrust of his project.[16] Technically speaking, theology (*theologia*)—which modern interpreters might call "the doctrine of God"—plays a very small role in Gregory's thought. As one can say nothing positive about God's being (*ousia*) and left only to speak of his activity (*energeia*) in the world, Gregory's primary concern seems to be with the philosophical life.[17]

Therefore, the distinction made between being (*ousia*) and activity (*energeia*), which is frequently taken to be a central principle of Gregory's metaphysical system rather than serving as the basis for a speculative theology, could be read as short-circuiting the understanding of this distinction as an ordering binary. Perhaps that which at first glance seems to separate the divine from creation does not give us a God's-eye perspective but serves to remind us of our own limitations. Perhaps the distinction does not provide one with a timeless fact but serves to begin to move one away from overestimating the representational capacities of human thought to know God as an object or a being, and shifts instead into a mode of theology that embraces its creative capacity and orients its discourse toward virtue and transformation. Perhaps the distinction is not a logical one that gives us information about things that do not change but has a performative force that draws attention to human creativity by way of noting the contingent character of the distinction itself. Gregory insists that since "no title has been discovered to embrace the divine Nature," we are left with the task "to name the Divinity, as we hunt amid the pluriform variety of terms applying to him for sparks to light up our understanding of the object of our quest."[18] Giving our breath to these sparks and kindling the flames of these words, one can begin to harness their transformative energy. Doing so, however, does not plug one into the depth of the divine being, as we are not given direct access to anything. Thought is "the cooperation of our own intelligence in interpreting the things that are done," not the correspondence of our formulations of truth to reality.[19]

Gregory's sixth sermon on the beatitudes, though it is often interpreted in metaphysical terms,[20] illustrates the performative role of the *ousia/energeia* distinction. The sermon is centered on the classically confusing beatitude "Blessed are the pure in heart, for they shall see God."[21] This Christic saying seems to inspire questions rather than directly conferring the vision that it promises, because it contradicts a number of other biblical passages that declare any vision of God to be simply impossible.[22] Rather than providing its hearer with firm footing, Gregory describes the situation of one who reads this beatitude to be like one who stands upon the precipice of a coastal mountain that has been worn away by the waves of the insurmountable deep. Accordingly, the beatitude does not grant assurance but is like that "slippery and precipitous rock which offers no foothold for rational thought upon itself."[23] He couples this sea imagery with the Matthean account of Jesus's walking upon the water, in which the frightened Peter is able to walk upon the waves when Jesus extends his hand out to him.[24] He writes that "the hand of the Word reaches out also to us and, as we are swept off our feet by the flood of speculations, sets us firmly on stable thinking."[25]

At first glance the assurance that comes from the assistance of Christ seems to undermine every seemingly slippery surface that Gregory has just warned us about with a paradoxical "firm and solid water."[26] In this context one might imagine that faith is that which overcomes the epistemological gaps of any merely human mode of knowing with the supernatural facts of revelation. For some readers, this is precisely what Gregory is doing, as he solves the dilemma of the beatitude by noting that our vision of God is neither the vision or knowledge of God's being (*ousia*) or nature (*physis*), but it is the knowledge that comes from seeing God through his creation and his activities in the world. Thus, one can know God by analogy as one can come to know the craftsman of an object through his work.[27]

To arrive at such a reading, however, one would have to cease listening or reading midway through the sermon, as Gregory proceeds to write: "The meaning of the Beatitude does not relate only to the fact that it is possible to draw an analogy from an operation to the operator. Even the wise of the world (1 Cor. 2:6) might from the harmonious structure of the universe arrive at an understanding of the transcendent wisdom and power. There is something else which in my view the grandeur of the Beatitude suggests to those able to receive its counsel."[28] Rather than imparting a largely abstract insight into the underlying structure of knowledge, Gregory is arguing for a performance, a transformation of activity in the one who tries to find his or her footing on this slippery rock or walk on the water of this elusive sentence. The counsel, then, that the beatitude offers, is similar to the good of health, which, he says, is not knowing "what health means but living a healthy life."[29]

Accordingly, what follows from this beatitude is not a proposition but an exercise, an *askesis* that forces one off the solid ground of rationalism or beyond the supposed givenness and calculability of sheer materiality. The difficulty of this thought causes one to see the world as shot through with the impossible light of the divine. This light does not simply reveal an intelligible object, but "with the materialistic fog removed from the eye of the soul, in the pure shining of the heart you see clearly the blessed sight."[30] Such a sight seems not to be entirely different from the activity of the seeing, as its vision is not of the being of God nor of the intelligible/sensible binary of a metaphysical fact. This sight does not give a theoretical and knowing subject *information* about an existence that it views from a distance. It demands exercises that lead to the *transformation* of one's life.[31]

## From Theo-logic to Theo-poetics

Read in light of this performative pedagogical, spiritual, and philosophical movement, the seemingly metaphysical distinction between God's being (*ousia*) and activity (*energeia*) takes on a different valence. The metaphysical preoccupation with being and substance and the rigid linguistic task that accompanies their philosophical primacy are replaced by a more dynamic theological concern with activities, with events.[32] This distinction, in its very theoretical content, undermines the division between theory and practice. What it is giving—a seemingly fixed distinction that can be possessed by a knowing subject—is precisely what it is taking away. The being and nature that Gregory allows himself to speak of never arrives into his possession but instead remains always absent, exceeding his grasp and demanding virtuous activity. In so doing, one is led not into a static and given domain of rightly ordered terms, but into an ongoing practice of interpretation and negotiation in which one struggles to find the sparks and glimmers among dynamic events.

The mode of writing theology that issues from this emphasis upon activity displaces the reductive parameters of metaphysics that some, like the early Derrida, have often associated with "theology." A speculatively oriented theology would undermine the dynamism of the *ousia/energeia* distinction, as these terms would be *"comprehended . . .* within the totality and enveloped in a volume or a book." In so doing, *written* theology would be protected from the drift of signifiers, as they merely bore witness to an underlying, unified, and coherent theory. Derrida characterizes this as "the encyclopedic protection of theology and logocentrism against the disruption of writing, against its aphoristic *energy.*"[33] The power of *energeia*, however, is the force with which it both calls for the positive work of writing and undermines the expectation of its ever having arrived at a static formulation. Gregory's theology seems to be oriented more toward the transformation that comes from the aphoristic energy of poetic invocation than the project of proper representation. Rather than employing terms with mathematical precision and thereby gaining access to the calculable essence of things, his theological discourse assists its reader with seeing the theophanies of the divine activity in the world with the wonder that comes from the perplexity of mystery. In so doing, the comportment of the theologian shifts from a position of control and possession—in which one stands above one's work, language, and creation—into one of creativity and struggle driven by humility and reverence.

To begin to come to terms with the simultaneous creativity and contingency of any such discourse and activity, however, is to also begin to undergo a shift of thinking from logic to poetics. Whereas logic is concerned with the proper governing and instantiation of entities, organized in the binary of what is actual and what is possible, poetics deals with the indeterminate and excessive occurrences and actions that exceed our grasp.[34]

The distinction between logic and poetics should not be taken as too rigid, however, because the basis of poetics would then be fundamentally secured by a logic. Accordingly, just as one cannot quarantine the poetic and creative dimensions of thought from logic, one cannot insulate the poetic from the logical. Yet, so long as logic remains obliged to dance with poetics, its primary intention (as it seems no longer bound by correspondence) and mood (as one is no longer stuck in the indicative) are radically changed. A theo-poetics, then, that sought to address itself to energy, to God's activity in the world, seemingly would exceed the task of representation or proper formulation and would attempt to undergo multiple performances and gestures. John D. Caputo writes, "As a symbolic discourse, then, a poetics is a certain constellation of idioms, strategies, stories, arguments, tropes, paradigms, and metaphors—a style and a tone, as well as a grammar and a vocabulary, all of which, collectively, like a great army on the move, is aimed at gaining some ground and making some point."[35]

Yet such creative activity would not merely address itself to these theophanies, or address its experiences of such to its audience. First it would be called from without, it would be addressed. To be constituted from without and caught up in the force and flow of that which addresses the theologian would seem to also announce the risk and promise of such a manner of speaking. Following Friedrich Nietzsche, one might acknowledge that all of the different domains and modes that Caputo has listed are to be positively deployed, but that the theologian, as a creative "subject," might not be steering the theological ship. That is, one should be wary of the ancient mistake that focuses on the *energy* that comes from the helmsman's imagined goal as he steers and forgets the driving *energy* of the steam that is propelling the boat or an even more powerful current in which it may be caught.[36] In fact, these other currents and energies may be so strong as to force the helmsman's hand, to steer the wheel and move his body, just as the drives and forces that constitute our own will could be much more complex than we could even begin to imagine.[37]

If theopoetics is not merely a tool or instrument through which one guides traditions and idioms to one's own ends but is also addressed by them, it follows that the poetry of thought cannot be made entirely secondary to or derivative of a guiding domain. The emphasis placed upon transformation and practice over and against a speculative and theoretical understanding of Gregory's theology is not a mere reversal of the theory/practice divide, in which all theory is taken to be derivative of a primary and "real" practice. The movement toward a poetic mode of theology has not merely replaced the solid intelligible referent of Gregory's "mythical" idiom with a sensible referent. The interpretation of his theology that I have been outlining does not reduce its images, idioms, and gestures to a real and practical end. Instead it finds itself caught up in the play of tradition and creation, invocation and interpretation.

Gregory's own poetic allusions are not arbitrary, nor are they simply an illustration of a larger philosophical point. The metaphors of "energy," "theophany," and "light" (which I focus on below) are not neutrally deployed to gain an intelligible *or* practical insight. Instead, these tropes should be seen as fitting into a larger tradition of thought and therefore not simply as examples. The transformation of the sun, light, and vision shows both a radical transformation of thinking and activity. This coupling of image and idea performatively subverts and creatively employs theology and philosophy so as to transform one's comportment toward the world. This transformation does not take place within the easily settled domains of either the intelligible or the sensible, but in the theopoetic play of the undecidable space between them. Gregory's theopoetic use of light employs an aphoristic and poetic energy that invokes, transforms, and creates as it obliquely eludes our grasp. In the play of light and the images that Gregory employs, or that employ Gregory, to show us the divine activity in the world, our "mind's eye" is trained as it begins to change the very way we see. In this play the binary between an intelligible and practical optics blends so as to give us a dynamic and reverent disposition.

## Gregory's Performative Optics and the Theopoetics of Light

In Western philosophical and theological traditions, light is often associated with a dualistic vision of the human capacity for knowing. Whether it is used as an image for the intelligible world outside the sensible shadows of the cave or as the Promethean fire that is stolen from religious superstition so as to enlighten modern humanity, light is often associated with knowledge and opposed to the darkness of ignorance. This opposition in the above-mentioned cases links the affirmation of light to the denial of another domain. In the former, light is opposed to the darkness of the sensible earth, and in the latter it is opposed to a sense of religious reverence and mystery. Gregory of Nyssa's use of the imagery of light differs from those traditions in that it is not stabilized by a simple dualism but is deployed dynamically. As such, its place in his thinking serves to overcome a speculative rejection of the sensible or an overly controlling disposition toward nature.

While Gregory largely deploys light with questions concerning knowledge and vision (particularly of the divine), light is not always associated with a positive form of knowing. His focus is on the disposition and activity of those upon which light shines. In his dispute with Eunomius, Gregory deploys the imagery of light both negatively and positively. At one point in the debate, in which Gregory has stated that he does not think our "perceptive faculties" can even come to know "the lower creation," he questions the boldness with which Eunomius speaks of God. As part of the rhetoric of his polemic, Gregory aligns Eunomius's epistemological optimism with the naiveté of childhood. He writes that Eunomius's discourse

> is like infants, who may be seen in the ignorance of childhood playing
> and quite serious at the same time. Often, when a sunbeam streams
> upon them through a window, they are delighted by its beauty and
> pounce on what they see, and try to take the sunbeam in their hand,
> and compete with each other, and grasp the light, catching the ray,
> as they suppose, in clasped fingers; but when the clasped fingers are
> opened, the handful of sunbeam makes the children laugh and clap
> because it has slipped from their hands.[38]

This condemnation of Eunomius's way of thinking as being childish is made complete as he compares these children to those whom Jesus rebukes in the parable of the marketplace.[39]

Though one might presume that the biting tone of these statements means Gregory opposes his own manner of thinking with the "metaphors" of light and childishness, later, in this same text and speaking of the same epistemological difficulty, Gregory employs the light of the sun in a positive sense. He does so in a rather traditional way, as the sun is said to represent God as that by which things become visible but which itself is invisible. Yet Gregory couples this rather traditional deployment of the sun with another analogy: though God's nature remains "unapproachable for immediate presence," the rays of the sun are like "a compassionate mother joining in the baby-talk with the inarticulate whimpering of her babies." It is in this manner, then, that the divine "passes on to the human race that which they are capable of receiving."[40] With this image Gregory seems to make it clear that all of our musings, and not just Eunomius's, could be characterized as childish. Furthermore, in so doing we can see that Gregory is not giving to himself precisely what he has taken away from Eunomius—a metaphysical schema whereby everything is properly organized. By coupling these two sun-drenched images, Gregory characterizes all thought in terms of childlike playfulness. One could perhaps imagine that theology is itself an activity of grasping at sunbeams.

What is different between Gregory's and Eunomius's approaches, however, is the expectation of the result such an activity might yield. Gregory says as much as he condemns this childish play by Eunomius, because he and those like him "see the divine power illuminating their minds through the words of providence and the wonder in creation, like the radiance and warmth issuing from the physical sun; yet rather than marveling at the divine generosity, and revering the one thereby made known, they overstep the mind's limitations and clutch with logical tricks at the intangible to catch it."[41] The light of creation, therefore, does not lend itself to being translated into a purely intelligible and logical domain. This light is not simply defined as the medium of knowledge, but its energy holds the paradoxical force of the in-breaking of God into our world, and therefore images are theophanies and not merely objects.

Theophany, here, seems to invoke a manner of acting and thinking that smiles, plays, and babbles in wonder and reverence. In this context, then, for Gregory, the light that illumines creation is not that which allows us to take the position of the knowing subject who stands apart from the world, seeking to control it through the proper identification of objects. This light is not the light of possession and calculation that would come from a divine light that serves as the ground upon which one would build one's system or tear the veil away from nature. Theophany is not the intermediary through whom we are able to make our conceptions correspond to the realm of truth. It could be thought of, instead, as the slippery sunbeam whose warmth we may feel and whose beauty may evoke wonder and reverence rather than the revelation of metaphysical fact.

The dynamic and pedagogical character of Gregory's use of light is perhaps most fully opened in a procession of theophanies in *The Life of Moses*. When Gregory treats the theophany of the burning bush, he does so in what seem to be largely rationalistic and static metaphysical terms. Gregory seems to posit that through this theophany Moses was taught that nothing subsists through what is apprehended by sense perception, but that "the transcendent essence and cause of the universe, on which everything depends, alone subsists."[42] Here, too, however, the function of light remains slippery as this initial image is overcome by an even higher insight. In the itinerary of *The Life of Moses*, the clear and direct apprehension that comes from the burning bush is displaced by the darkness of a cloud that covers the mountain where Moses communes with God. The transcendence of God cannot be directly or intellectually apprehended.

In the cloud, the essence of God is displaced as it both blocks and exceeds our vision.

In regard to God's nature, the human capacity to know is largely understood in negative terms. The apprehension of "true being" is not seen clearly, but rather it is "the seeing that consists in not seeing."[43] The mode of thought that attempts to reduce God to being a concept or a mere image is displaced by a negative vision, by the present absence of "luminous darkness." After progressing through this delimitation of representational thinking (through the burning bush and the darkness on the mountain), Gregory asserts the primacy of a performative orientation as he frames the epistemological negativity of thinking within the larger scheme of religious virtue. He divides this itinerary for thinking into two parts. Gregory writes, "Moses learns at first the things which must be known about God (namely, that none of those things known by human comprehension is to be ascribed to him). Then he is taught the other side of virtue, learning by what pursuits the virtuous life is perfected."[44] Accordingly, much like the sixth sermon on the beatitudes, though God's nature is said to be unknowable and invisible, what is emphasized is the mode of acting in the world that follows from the darkness of knowledge. Rather than affirming an alternative metaphysical system, the vision of theophany leads to a performative space.

The movement of *The Life of Moses* does not lead into a directly prescriptive discourse on how one should live one's life. Rather than providing a code or a rule, it employs another biblical image: the tabernacle, the holy of holies.[45] Whereas the cloud of darkness emphasized the transcendence of a God that is outside our comprehension, the holy of holies is an image that reverses such a relationship, emphasizing God's immanence in creation. As Scot Douglass points out, here the outside (God) goes inside, into the holy of holies, and the inside (human) is forced outside. In this image the divine is not out of reach on a mountain but is present within the inner sanctum of the temple. However, this does not mean the outside has been appropriated by the inside and made into an object. When God is present in the inner sanctum, no one is able to come into this space, not Moses, not Solomon, not anyone.[46] That is, except for Christ, who is the heavenly tabernacle. He is that which is unmade but made among us.[47] This assertion of the incarnation in this imagery is not the deployment of a metaphysical hinge whereby the paradoxical is made rational, but it instead demands a dynamic understanding of names and images.[48] The holy of holies unsettles rather than orients, as the darkness of knowledge regarding God does not allow for a static or rigid disposition toward that which is seen in light.

This image does not lead to a deeper intelligible fact, in which the sensible is merely made derivative. The incarnation of God in the world creates a play of undecidability that demands both our creativity and humility.

The line of thought that Gregory seems to be trying to open does not merely use these figures and activities as vehicles to arrive at an intelligible and representational truth. His theopoetics seems to be weaving them together as different elements in a textual exercise. Read in this way, *The Life of Moses* is the imaginative deployment of ancient biblical images for the inspiration of a philosophical life. Therefore, in contrast to many of the dominant accounts of the "Greek" translation of Christianity in Late Antiquity, Gregory should not be read as assimilating the sensible Hebrew scriptures to a normative, intelligible Hellenistic framework; instead, his work haunts the intellectual and cultural framework of his world by negotiating with an alien and elusive tradition. Douglass notes that this tactic led Gregory to look for the meaning of a text between its letters, in the spacing of language, and in the tension with tradition and not just in its underlying referent. Douglass writes that by "inserting himself in the gaps between the life of Moses" Gregory sought to bring out "the implications of the narrative concerning Moses and the lives of fourth century believers."[49]

To view or build the tabernacle, then, is not to grasp in one's apprehension the beam of sunlight, but to be filled with reverence and wonder. Though this does lead Gregory, in the formerly noted instance, to place us in the position of children, it does not undermine our ability to speak or our capacity to create. We can still create and manipulate things in our world. Gregory is challenging the pretense that the intellectual *energy* that we harness from the sun solidifies into a rigid *actuality*. This shift in thinking demands that we look at our creations as fleeting and transformable. The holy of holies and the incarnation do not serve as the ground upon which one can organize all of creation but as the glimmering light that illuminates the crevices of our creation and demands our reverent activity. These images, therefore, come to have a performative force that transforms how one looks at and acts in the world.

Gregory notes that these images lead to activity, as when one looks to "the tabernacle above, he sees the heavenly realities through these symbols"; but when he looks below he sees the servants who are the pillars of the Church. These pillars are not described as the bishops of the church that enact the holy of holies through their sacramental functions. He expands these pillars to be "Peter and John and James" *and* "all those *who* themselves support the Church and become lights through their own works."[50] One obtains the priestly garments of light through the joining of both contemplative and practical philosophy.[51] From the imaginative activity of contemplating the dynamism of the symbols above and the ascetic work of the virtuous life below, one changes one's relationship to creation. Gregory writes that the priestly figure "should not inflict upon his soul a heavy and fleshy garment of life, but by the purity of his life he should make all the purists of life as thin as the thread of a spider web. Reweaving this bodily nature, we should be close to what rises upwards and is light and airy."[52]

This creative activity of reweaving our tunics into priestly vestments seems tied up with a disposition that undermines the intelligible/sensible binary, and should not be read as a simple rejection of the body or this life for the sake of an intelligible actualization. The pillars of the church overcome the vices of possession and the hubris of conceptual idolatry, and not bodily life all together. Through this activity our bodily existence gains a lightness that no longer requires the arduous labor of false expectations. Though these tunics are made light and almost ephemeral by ongoing and never-ending exercises, the work of the philosopher, like the Israelites out in the desert, is to cease the game of possession and hoarding and to instead merely gather the minimum amount of the manna from heaven that one needs and devote our treasures to the forming of the Tabernacle. Just as the burning bush is not consumed by fire to become pure light, the vision that comes from theophany does not lead one beyond the material world. The light of theophany, instead, leads one beyond a perspective that sees things as static and easily possessed objects and into the opening of the divine, which inspires reverence, wonder, and compassion.

The theophany that Moses receives on the mountain changes his very outlook, the basic position of his optics. He does not, as an overly theoretical position might imagine, obtain a cartographic view of the lay of the land from on high, nor does he come to see God's face, as though he could fit these into the metaphysical frame of a direct view. Gregory emphasizes that Moses was able to view only God's back.[53] In so doing, he contrasts a way of seeing that is built on the static view of standing apart and perceiving an object with the active movement of following.[54] It is impossible to see God and live, because God is the way, the truth, and the life, and the only way to truly live is to follow his path. Therefore, if one were to take the overly rational, metaphysical, and idolatrous position that attempts to face God (the optics of the subject that stands apart from and represents its object), one would not find oneself in a direct face-to-face relation with the divine, but would find instead that in this position one's "course will certainly be in the opposite direction."[55]

This shift in optics does not preclude human creativity and ingenuity in favor of an entirely passive resignation but redirects it. Gregory merely wants instrumental thought to be oriented toward a disposition of reverence and simplicity. He notes that the same conceptual creativity (*epionia*) that his theological discourse relies upon is what drives the activities of geometry, logic, arithmetic, and mechanics. He goes so far as to write that "every beneficial thing useful to mankind which time has invented was not found out except by the use of conceptual thought."[56] What is central to his way of thinking is the degree to which these activities are undertaken under the pretense that the sunbeams of creation may be possessed, that the theophany of God's work in the world is a sheer materiality that is standing in reserve to suit our whims and desires. The energy of our thought holds both promise and risk, as "the one who steers the ship into harbor with the rudder could also steer it onto reefs and promontories."[57]

Energy itself holds both promise and risk, as it opens new creative vistas for our lives, while it also threatens to eradicate the very means of life itself on our planet. So long as we remain deaf to the theopoetic call of alternative energies, we may not be able to change course from the destructive currents of our Promethean outlook. If, however, we can begin to hear the melody of a different account of our knowing, if we can rescore the visions of light, then perhaps the activities of our lives can become redirected and simplified. Perhaps the choruses of greed and fear that invoke the energies of our time can be transformed by an alternative optics that does not see itself as possessing the light and power of the fire of nature. Perhaps instead we may begin to live songs of praise that sing out of wonder and simplicity. In this manner theology and philosophy might, from the exercises and creativity of our lives, help to provide an alternative energy that is reverent and sustainable.

# CHAPTER 4

# Beyond Heat:
# Energy for Life

## Clayton Crockett

In his book *Theology of Money*, British philosopher of religion Philip Good-child claims that "theology consists in the ordering of time, attention and devotion" in a broad sense rather than the determinate faith in Jesus Christ or any other particular religious tradition.[1] If God becomes questionable in the modern world, largely as a result of scientific discoveries and developments, then theology either disappears or becomes transformed, unless it simply digs in its heels in a reactionary way. I suggest, following Good-child, that theology involves the ordering of time, attention, and devotion in relation to what matters most, what is of supreme value. For many modern humans, this most cherished value is life. And for me, then, loosely following Goodchild, theology involves conscious reflection about and evaluation of the value of life. Goodchild thus argues that "the divorce between the secular and the religious, between attending to treasure on earth and attending to treasure in heaven, must be overcome."[2] We cannot simply posit and oppose an absolute, transcendent value elsewhere that denigrates life here on earth, and we diminish our experience of living by trivializing it and prohibiting its valuation in religious and theological terms.

I argue that the essence of life in a material sense is energy conversion, and that it is our use of energy and energy resources that allows humans to achieve such incredible standards of living in material terms that they enable cultural expressions of human value and civilization. This discussion presupposes human consciousness as an important function and value, but it does not privilege consciousness over against other forms of life. In this essay I briefly survey energy exploitation over the course of human history and then engage with some of the findings of nineteenth- and twentieth-century physics in order to suggest that we need to think about energy in non-thermodynamic ways. This a-thermal perspective will result in a somewhat radical and hypothetical proposal, based upon the research of Kevin Mequet, an architect who has synthesized broad aspects of energy, geology, cosmology, and theoretical and practical physics. I return to theology at the conclusion and suggest a materialist theological perspective that is neither atomistic nor crudely reductionistic.

Solar energy supports life on earth. The main source of energy conversion is photosynthesis, where plants combine carbon dioxide with water and photons to produce carbohydrates and oxygen. The photons supply the energy necessary for the conversion to occur.[3] A carbohydrate is a store of energy and a source of food or fuel for plants and, secondarily, animals. At a basic primary level, then, energy conversion is not about heat, even though heat (in the form of plant metabolism) is a result of this process. Animals cannot directly convert sunlight into energy; they must consume plant carbohydrates directly or indirectly, by consuming other animals. Animals convert carbohydrates into energy by means of oxidation: they burn fuel and produce heat. "Most of the energy released from food by an animal's metabolism immediately becomes heat," writes biologist R. McNeill Alexander, "but some may be used to do work, and so converted to other forms of energy."[4] Animals primarily produce heat by burning up food, and it is this process of energy conversion that allows them to live and to labor.

This animal model of energy conversion informs all of our human efforts to exploit and utilize energy, culminating in the extraordinary civilization produced over the last three centuries. Basically we find external sources of fuel to burn, and this burning produces surplus energy to use for material, agricultural, structural, and later cultural endeavors, but these processes of energy conversion also release extraordinary amounts of excess heat. The domestication of animals and plants has allowed humans to put other forms of life to work for their ends, and this domestication that makes possible human culture is predicated upon the domestication of

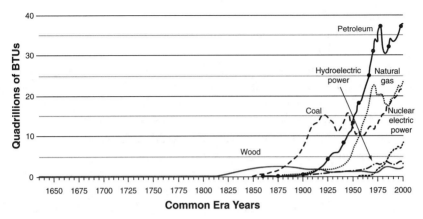

USA energy consumption by source, 1635–2000 C.E.

fire. The anthropologist J. C. Heesterman states: "not only were selected animal and plant species domesticated; most centrally, and first of all, fire was domesticated."[5] The ability to burn wood and other plant and animal products enabled humans to develop what we call human culture and civilization.

As with most forms of evolution on earth, the history of energy exploitation shows a long, slow, seemingly static development that then takes off exponentially as we approach the present. A chart beginning in 1635 C.E. shows for the United States what we could call the "energy grand staircase," the development of newer and more productive sources of energy.[6] Until the nineteenth century (the eighteenth century in England, where the industrial revolution began), the primary source of energy was wood, which was supplemented by domesticated animals, the enslavement of the energy and labor of other humans, and small amounts of energy derived from water and wind.

What is extraordinary, as the chart shows, is the enormous quantitative change marked by the transition from wood to coal, and then another huge transition from coal to petroleum and natural gas. The industrial revolution occurred because of the ability to extract and use new sources of energy, fossil fuels, that could be burned for much higher rates of return. Relatively enormous amounts of relatively cheap energy produced the contemporary civilization that most of us now take for granted, at least in wealthy parts of the world. And this civilization would not be possible were it not for the extremely high energy return on energy invested (EROEI) of oil.

All of this sounds really impressive, but the dark underside is that these sources of enormous energy are nonrenewable, at least on timescales relevant to human life, and we are approaching or have reached a limit in terms of the amount of oil that we can extract and produce over a finite amount of time. We are quickly consuming "the last hours of ancient sunlight," to use Thom Hartmann's phrase.[7] In 1956 Shell Oil Company geologist M. K. Hubbert predicted that U.S. domestic oil production would peak in 1970, and world oil production would peak and then gradually decline around 2000.[8] Hubbert was correct about U.S. oil production, and a result of this occurrence was the abandoning of the gold standard in 1971, followed by the OPEC (it was called OAPEC [Organization of Arab Petroleum Exporting Countries] at the time) "oil-shocks" in the early to mid-1970s. As a result, the United States reaffirmed its political and military support of the corrupt regime of Saudi Arabia, by far the largest oil-producing nation, and the Saudis were able to flood the market with oil throughout much of the 1980s and 1990s. For a number of reasons, including the Iranian revolution in 1979 and the U.S. economic recession in the early 1980s, the peaking of world oil production was delayed until the first decade of the twenty-first century, but since 2005 oil production has been running along a plateau, and many countries are already facing serious declines.[9] There are many critics of Hubbert and deniers of peak oil, but regardless of the exact timing, the scenario of a current or approaching peak seems undeniable. We also see many attempts to exploit alternative, non–fossil fuel sources of energy, but unfortunately none of the available alternatives currently possess anywhere close to the EROEI that oil does.

Furthermore, as we desperately search for more and more fossil fuels to burn for energy to fuel our growth-dependent economy, we release more and more heat and heat-trapping gasses like carbon dioxide and methane into the atmosphere, which threatens to heat the planet to temperatures that are not compatible with human life in the comfortable forms to which many of us have become accustomed. The Intergovernmental Panel on Climate Change (IPCC) estimated in 2001 "that global temperatures will rise by between 1.4 and 5.8 degrees [Fahrenheit] this century."[10] As George Monbiot points out, a rise of global temperature of just 2 degrees Fahrenheit will have devastating impacts upon sea levels, food production, and species decline and extinction. Furthermore, some climate scientists believe that "the temperature this century could rise much further" than the IPCC's estimates.[11]

Peak oil and global warming are two interrelated expressions of one fundamental problem: we have exceeded the capacity of the natural resources of the earth to sustain so many billion people at the levels of consumption that we have reached, and that many developing countries are reaching, in physical terms. Contemporary corporate capitalism, which serves as an ideology and even a religion for many people, especially in the first world, is based upon a lie, which is the fantasy of indefinite if not infinite growth.[12] Global capitalist growth was uneven, unfair and poorly distributed, but it was possible because of extremely cheap forms of energy, and now energy is becoming scarcer and more expensive. Reality conflicts with capitalist assumptions and projections, because we cannot infinitely grow with a finite resource base. We are reaching the limits of human ability to expand and manipulate our environment, even as we desperately search for new technological breakthroughs that would allow us to continue doing so.

It may not be possible, and it may be too late, but I suggest a reorientation in thinking about energy and energy conversion, one that is not based upon heat. Even if such a reorientation does not lead to practical innovations, or if such innovations prove to be too little and too late, I still believe that this fundamental turn away from heat may possess theological, philosophical, and ethical value. Going back to plants, we need to think about energy more directly in terms of sunlight, and not in terms of heat and burning. Solar energy is electromagnetic, and energy can be viewed in terms of electromagnetic conversion, even if nuclear forces originally give rise to this energy. The earth has its own energy source as well, grounded in nuclear fission, as we will see, but let's focus on electromagnetism directly for the moment.

Building on the experiments of Michael Faraday, it was James Clerk Maxwell who united the newly discovered forces of electricity and magnetism into one by means of mathematical equations. Maxwell, following Faraday, saw that "an oscillating magnetic field would give rise to an oscillating magnetic field . . . and this oscillating electric field would, in turn, give rise to an oscillating magnetic field."[13] So a magnetic field produces an electric field, which produces a magnetic field and so on. What Maxwell did that was so incredible, however, was to work out experimentally that this propagation must occur at the speed of light! Maxwell's result led Einstein in 1905 to derive the speed of light mathematically from electromagnetic energy conversion. And it was the significance of the speed of light for electromagnetic fields that led Einstein to posit that speed as a constant limit, and to vary everything else, including time and space, which led to the most important breakthrough in twentieth-century physics: special relativity.[14]

Electromagnetic energy conversion occurs at the speed of light, and this process also applies to the conversion of mass and energy. That is, mass is simply a very special form of energy, an insight (along with some brutally complicated mathematical equations) that leads to the development of general relativity.[15] General relativity extends the relativity of space and time to gravitational phenomena, where gravity warps space-time in relativistic ways. Matter is this distortion of space-time affected by gravity. Matter is not primarily physical stuff, but it is a highly condensed mass of energy. Atomic physics, which Einstein opened up by conceiving light particles as quanta but then partially abandoned, involves breaking apart the atom in order to release vast amounts of energy. Although physicists later in the twentieth century became obsessed with the search for elementary particles, in some ways the very notion of a particle becomes incoherent at the subatomic or quantum level. As Werner Heisenberg puts it, "We have learned that energy becomes matter when it takes the form of elementary particles." Ultimately, however, "every particle consists of every other particle," which derails the hopes for a consistent atomic reduction to the smallest constituents.[16] Matter is best understood as a concentrated form of energy, and a particle is more often a virtual particle that "blinks" into and out of existence rather than a "real," stable one.

What do we do with this enormous amount of energy that is contracted in matter? How do we exploit it? Keep in mind that heat is not the essence of electromagnetic conversion, or of the energetic nature of photons and electrons, although they do produce heat when excited. Heat also provides a measure of entropy, or increasing disorganization, which is the only way physicists have to mark the flow of time from past to future states.[17] Entropy is a loss of organization, which ultimately leads to death. Energy is lost as heat. Most human activities are neg-entropic (negative entropy) in the sense that they produce rather than dissipate organization and information, but only in very special and limited ways. Ultimately and necessarily, according to the laws of thermodynamics, more energy is lost as heat than is utilized for energetic, productive work. Heat energy, the method by which we naturally and normally use energy, is largely waste, and it tends toward death.

Nineteenth-century conceptions of entropy have dominated many twentieth-century scientific and philosophical understandings about the nature of life, but in many ways this view of entropy as the inexorable law or fate of the universe is an illusion. As Gilles Deleuze argues in *Differ-*

*ence and Repetition*, in championing entropy we absolutize a partial truth. Entropy concerns the extensity of energy, how it is expressed and applied, according to the principles of thermodynamics worked out over the course of the 1800s, beginning with Sadi Carnot. According to Deleuze:

> The paradox of entropy is the following: entropy is an extensive factor but, unlike all other extensive factors, it is an extension of "explication" which is implicated as such in intensity, which does not exist outside the implication or except as implicated, and this is because it has the function of *making possible* the general movement by which that which is implicated explicates itself or is extended. There is thus a transcendental illusion essentially tied to the *qualitas*, Heat, and to the extension, Entropy.[18]

In other words, entropy is the measure of an increase in heat and a decrease in organization, and it is what is being measured. Entropy seems to explain both the intensity and the extensity or extension of energy, which is a problem for Deleuze. Deleuze suggests, on the other hand, that we understand intensive energy along the lines of his philosophy of difference, where intensity drives difference and difference expresses itself by repeating itself, differently.

Thermodynamics reduces differences; it burns them up in fire. A thermodynamic approach "goes from the side of things to the side of fire: from differences produced to differences reduced."[19] Deleuze wants to preserve the integrity of intensive difference, which resists any simple entropic reduction. He does not explicitly use the language of quantum physics, but we could say that quantum electromagnetic energy is intensive energy as opposed to thermodynamic extensive energy. The expression of energy as we measure it, at least at macro-atomic levels, is entropic, but the essence of energy itself is intensive, electromagnetic, which generates energy and produces differences. In *Difference and Repetition*, Deleuze opens up a theoretical perspective beyond thermodynamics, even if it is not fully developed in terms of twentieth-century physics. Intensive difference does not overcome entropy, but it is hidden by and operates underneath extensive thermodynamic processes. The second law is not wrong, but it is easy to absolutize in a superficial way, to misunderstand and misapply it.

The intrinsic conception of nuclear energy is non-thermodynamic, but we have been able to actualize it only in thermodynamic terms. Nuclear energy as it is currently conceived is an incredible force, but we have been unable to make use of it in comparison with fossil fuels. Notice how far below coal, natural gas, and oil the level of nuclear electric power is on the chart described above. All we know how to do is to burn nuclear fuel rods to heat water to produce steam to turn a turbine to produce electricity. And the production of electricity is far more inefficient than fossil fuels, not to mention the radioactive toxic waste left behind. Oh, and we know how to make a bomb, to start a chain reaction and blow things up. Nuclear power does not produce carbon emissions, although in its current infrastructure it is highly dependent upon fossil fuels. It is, however, the only currently available energy resource that could be imagined to replace fossil fuels on a large scale, but it is still woefully underproductive as well as highly expensive and extremely dangerous.

I'm going to go out on a limb. I'm going to speculate, based upon some of Deleuze's ideas as well as Mequet's research, about a new way to think about nuclear energy and nuclear power, and this is based upon an unproven hypothesis about what is happening inside the earth.[20] But sometimes you have to take risks, and our situation is desperate enough that it's at least worth considering. I don't know whether or not it's right, or if it could work even if it is, but it is an a-thermal way to think about nuclear energy.

Why does the earth possess such a strong magnetic field? Earth's magnetic field is anomalously strong compared to the other rocky planets close to the sun. We do not exactly know, and according to the classical dynamo theory of electromagnetism there would need to be a strong electrical field at the core of the earth to generate this magnetic field. The problem is that it appears difficult to conceive of a significant source of electricity within the earth's mantle and core, and the high temperature at the core (above the Curie temperature) would seem to preclude the conduction of electricity. For the sun, and for the gas giants (Jupiter, Saturn, Uranus, and Neptune), we know that they have plasma in their mantles and cores, which is extremely conductive of electricity. But the earth's core does not contain plasma.

So what is going on within the core and mantle of the earth? If there is no known source of electricity, perhaps the magnetic field is being generated directly by means of nuclear fission reactions (fissile and fertile uranium and thorium) within the mantle and core of the earth. This process is accomplished by the movements of and within the mantle, including convection and Coriolis motions. Earth benefited from a large amount of heavy nuclear elements that settled at the right distance away from the sun when the solar accretion disk formed. The other rocky planets, including Mars, seem to have had a magnetic field in the past, but now they do not,[21] or they have extremely weak magnetic fields, and this may be due to the exhaustion of their nuclear elements.

The nuclear reactions themselves involve particles that come together and exchange a virtual quantum that also produces a strange magnetic current according to Richard P. Feynman and Murray Gell-Mann's "Theory of the Fermi Interaction" that explains radiologic decay.[22] In beta decay a released neutron decays into a proton, an electron, and a neutrino by means of strange subparticles, giving rise to a net magnetic effect. Mequet suggests that the Feynman/Gell-Mann radioelectromagnetic effect is what is generating the earth's intense magnetic field. If this hypothesis is correct, then the earth's magnetic field is produced by nuclear reactions that certainly involve heat, but this heat may be masking what is crucially going on—that is, the direct generation of a magnetic field by nuclear fission.

This idea is just a hypothesis, and it has not yet been scientifically validated. However, what possibilities open up if we conceive energy production a-thermally? Since a magnetic field produces electricity, if we could somehow model the earth's production of magnetism with a nuclear magnetic generator, we could induce electricity with such a device. Again, this idea is still very experimental, but it is worth considering and evaluating. The key theoretical insight is that electromagnetism produces energy directly, without requiring thermal heating. Heat is a by-product of this process, but it is not the core of the process. We need to think seriously about energy in both practical and theoretical terms beyond our obsession with heating and burning fuel in order to survive.

What does it mean to live? In one of his last essays, "Immanence: A Life," Deleuze writes: "We will say of pure immanence that it is A LIFE, and nothing more. It is not immanent to life, but the immanence that is in nothing else is itself a life."[23] What does this mean? *Immanence* and *life* are two terms that are very important for Deleuze's philosophy, and he equates them here at the end of his life. *Immanence* refers to the elaboration of a plane of immanence, a plane of consistency that makes sense by way of referring to what is here and now rather than that which is hidden or transcendent. For Deleuze, thinking and living occur along a plane of immanence, and this immanence distinguishes what is integral to thinking and life, because it is purely and simply available to it in a univocity of being. Immanence does not exclude or restrict transcendence, but it disqualifies it from sense as transcendence. That is, immanence is the necessary precondition for making sense in our life and in our ideas.

What does it mean to say that immanence is "A Life"? Deleuze defines life more broadly than simply in organic biological terms. *A Life* means the consistency of self-organization, the mark of complexity that constitutes immanence as pure immanence, an immanence that becomes itself by means of this process of self-organization. Energy can be seen as a process/plane of immanence in a way that disrupts the simple dichotomy between immanent and transcendent, material and spiritual, insofar as energy constitutes life, A Life. Energy is immanence, and pure immanence for Deleuze would then refer to the self-organization and transformation of energy conversion that constitutes life in its broadest sense. Energy is both spirit and matter, as well as neither simply spirit nor matter in the way we ordinarily understand those terms.

Energy is the precursor to life as we know and experience it, and we are an expression of energy organization and transformation. The form of energy conversion with which we are most familiar is thermodynamic, due to the biological nature of our being. We burn stuff in order to release energy for work, and we give off massive amounts of heat. This process is relatively inefficient, but it is the way we live, and for the most part we have modeled the production of energy as fuel and electricity based upon our biological systems. On the other hand, as Einstein has shown, there is another way to conceive energy production and transformation—that is, electromagnetically—and we still have not fully understood or applied the significance of his discoveries. If it is not already too late, we need to rethink energy in electromagnetic terms, beyond our obsession with heat, in order to continue to live on this planet.

Theology is most often understood as the intellectual defense of a prior faith commitment, usually to various forms of Christianity but sometimes also to Judaism or Islam. Faith is then seen very much in voluntaristic terms, the assumption being that we simply choose what we believe in and fully understand why we do so. The tradition of radical theology, after the death of God, questions these assumptions about what theology is and can be. The death-of-God theology is a twentieth-century movement in American theology, but it refers to an experience in the modern world of the absence or questionability of God.[24] Fundamentalism clings desperately to traditional understandings of God, while secularism follows the literal logic of the death of God and dismisses theology and religion in the name of nature or human culture and value. I understand theology much more broadly, as a mode of thinking and questioning about reality in an ultimate sense, without any pre-given answers.

If God as the repository of transcendent value no longer holds weight, then God may be thought as immanent to human and natural processes, even though this is a form of pantheism, or sometimes, following Whitehead and Hartshorne, as pan-en-theism. However, the most important task is not to redefine God. The most important thing is to know what functions as meaning and possesses value. Deleuze's philosophy of life is seen as a neo-vitalism, because it draws upon the vitalism of late nineteenth- and early twentieth-century thinkers such as Nietzsche and Bergson.[25] At the same time, Deleuze understands life much more broadly than organic life, and here he presses against the limit of Whitehead's philosophy of organism. How do you make of yourself a body without organs? How do you construct a way of living and thinking that avoids the *telos* that we usually ascribe to organs and organisms?

In this essay I have tried to suggest that we can think of life broadly in terms of energy transformation, and this energy transformation occurs along what Deleuze calls a plane of immanence. Immanence is a life, and energy charges it, brings it to life in a doubling of life. Theology as a discipline or form of thinking may attend to life in a materialist sense and think energy as spirit without the dualism of spirit and matter. Materialism here is neither consumerist materialism nor a reductive atomic materialism, but rather the acknowledgment of our implication in material reality, that life is the complex self-organization of flows of energy. A materialist theology, then, elaborates a vision of what it means to think and to live vitally in light of the using up of natural resources and energy reserves. We have to get used to physical scarcity, but that does not necessarily imply metaphysical poverty. At the same time, we must be careful that our metaphysics and our theology do not degenerate into simple wishful thinking or a fantastic idealism that coincides with the metaphysics of consumer capitalism. We cannot live without heat; heat is necessary for our lives. At the same time, we have to think about the limits of thermodynamics, the possibilities for life that exist in a-thermal ways. Entropy is the (second) law of thermodynamics, and entropy prescribes death to our efforts to live and self-organize and reproduce. But life and death are not simple opposites, even though for sexual beings they appear to be. If immanence is a life, that does not mean that immanence is God, but it does mean that the values to which we cling are not extinguished at the end of our physical life. Immanence continues as long as the universe. As does death.

# CHAPTER 5

# Emergence, Energy, and Openness:
# A Viable Agnostic Theology

## *Whitney Bauman*

> Organisms and all known functioning ecosystems—with the
> arguable exception of the biosphere itself—trade both matter
> and energy across their boundaries. They are open systems.
>
> ERIC D. SCHNEIDER AND DORION SAGAN, *Into the Cool*

> The new emerges far from equilibrium at the
> *edge* of chaos in a surprising moment of creative
> disruption that can be endlessly productive.
>
> MARK C. TAYLOR, *After God*

In their book, *Into the Cool*, Eric D. Schneider and Dorion Sagan argue
that the second law of thermodynamics, describing the increase of entropy
through energy exchange, is truthful, but not the whole truth.[1] Thermodynamics was tested and "created," so to speak, in the laboratory, in closed-
system experiments. In other words, it was discovered in a situation outside of the interconnectedness that organisms find themselves in outside
of the laboratory. For closed systems, entropy is the final word.[2] However, most living systems in the world (and perhaps the universe itself) are
open. "Our biosphere may not be an organism per se, but it is composed
of open systems—cells, populations, and ecosystems—whose energized
interactions make Earth's surface behave as a nonlinear, even physiologi-
cal entity."[3] These open systems do not operate in such a way that they
seek equilibrium. In fact, "for living beings, thermodynamic equilibrium is
equivalent to death."[4] Rather, they operate by "reducing energy gradients."
The more complex an organism, the more effective that organism is at
taking in energy and exporting entropy. "Dissipative structures grow more
complex by exporting—dissipating—entropy into their surroundings."[5]

Because living systems (whether a river, a cell, a gene, an organelle, a forest, or perhaps the earth itself) are open, energy and materials are constantly being exchanged in an effort to maintain a certain amount of disequilibrium between the organism and the environment that surrounds it. That is, all living systems are response-able. When an organism stops, becomes nonresponsive, and moves toward equilibrium with its surrounding environment—whether through consumption by another organism or absorption into the surrounding environment—death occurs. In other words, equilibrium means the end of being a response-able entity. At the same time, another organism gains the energy created by that death. For this reason, Schneider and Sagan call nonequilibrium thermodynamics (NET) the "science of creative destruction."[6] This implies that creation of new organisms and of more complex life-forms does not occur ex nihilo, but rather at the expense of other life forms. The "new" here arises at the expense of "others" and within evolving contexts. It does not come "from nothing," but emerges from everything.

This chapter takes the science of nonequilibrium to heart and mind. The maps of reality that much of our contemporary scientists create are no longer two-dimensional drawings of static materials in equilibrium or even harmony.[7] Rather, they are digitized 3-D maps of movement, providing us with moving, hypertexted scripts of a reality that is not in equilibrium, not in stasis, not even necessarily in harmony. Instead, from ecology to evolution to quantum physics, change is the name of the game. At the same time, this emergent, chaotic, nonlinear, and nonsystematic kind of thinking is very much a part of contemporary theological and philosophical discussion (as the opening epigraph from Mark C. Taylor suggests). Much of the recent and distant history of theological and philosophical reflection relies on some sort of foundational stasis (whether original or final). That is, an assumed equilibrium or foundation in the "natural world" has been reflected in the theologies and philosophies of these same types of sciences. These types of philosophies use transcendent foundations, forms, categories, and ends to order the worlds around us. However, "post" thinkers and theologians such as Catherine Keller, Gordon Kaufman, Ivone Gebara, J. Wentzel van Huyssteen, Mark C. Taylor, and others are challenging this foundational, static way of "doing" theology and, through processes of creative imaging, are developing some viable—that is, living and moving—theological maps of reality.[8] This work is a kindred attempt. Like the aforementioned theologians and for similar reasons, I want to tear down the theological dam(n)s of foundational ways of thinking: for these foundational dam(n)s block energy-thought flows, reify life, and lead toward many types of violence in the world. Furthermore, they create closed systems of thought, leading eventually toward creative entropy. This essay's addition to what I am calling a viable agnostic theology is a modified version of the content of theological reflection based upon an expanded understanding of *emergence* and Feuerbach's theological projection. In other words, I may end up in the same place as other post-foundational thinkers, but it is my hope that my somewhat different starting point will be a way of bringing "others" into an ongoing dialogue about our place(s) in this evolving planet.

Big hopes often start from humble beginnings, so I begin here with my own humble accounts of constructing a viable agnostic theology *of and for* the earth. In doing so I first look at some of the basic ideas behind emergence, then move toward a discussion of Feuerbach and epistemology, and finally end with the importance of agnosticism for creative-energy flows in a viable, evolving planet.

### Emergence: From Flesh to Word and Back Again

> Emergence theories presuppose that the once-
> popular project of complete explanatory reduction—
> that is, explaining all phenomena in the natural world
> in terms of the objects and laws of physics—
> is finally impossible.
>
> PHILIP CLAYTON, "Conceptual Foundations of
> Emergence Theory"

Rather than rehearse the history of emergence here,[9] I articulate three issues within the contemporary discussion of emergence that redefine *newness* and the relationship between thought and matter in a contextual way: autopoiesis, top-down and bottom-up causation, and multilayered reality. I use the term *autopoiesis* in its most literal sense: complex systems "auto-create." This might also be captured in terms of "property emergence" and the "irreducibility of emergence." "When aggregates of material particles attain an appropriate level of organizational complexity, genuinely novel properties emerge in these complex systems. . . . Emergent properties are irreducible to, and unpredictable from, the lower-level phenomena from which they emerge."[10] Autopoiesis, then, describes a context-dependent understanding of newness. These new emergents are neither dis-covered nor placed into existence by some dis-continuous, external force. They are neither reducible to "the same," nor are they created out of nothing. Rather, new things (much like new ideas) emerge from and move beyond specific bio-historical/natural-cultural contexts. These new emergents are not merely caused by, but also have causal effects on, the contexts from which they emerge.

Emergence theory provides the beginnings of a discussion on how to understand top-down causation as well as bottom-up causation. By "top-down" I mean that "higher-level entities causally affect their lower-level constituents."[11] Though many materialistic scientists might cringe at the suggestion, I would argue, along with other non-reductive scientists, that emergence theory might be a viable way forward through the Scylla of reductive monism and the Charybdis of Cartesian dualism. That is, I am arguing along with theologian Gordon Kaufman that our bio-historical projections have efficacy for what philosopher Donna Haraway describes as the natural-cultural worlds in which we live. Our "mental" events affect our brains, not just vice versa. Newness, then, comes at the levels of both biology and history. At times, our bio-historical projections—our God-talk, utopian ideals, and so on—can really affect the way the world becomes in such a way that new, complex relationships emerge. Granted, this newness does not always mean "goodness"; newness works both ways. The point is that both bottom-up and top-down causation (though neither should be confused with only direct causation) are necessary if there is to be newness as opposed to "more of the same." In a bottom-up-only explanation, everything is explainable from the same laws of physics. In a top-down explanation, everything (ultimately) is explainable by some external force—whether destiny set in the stars or an omni-God.

Finally, emergence theory suggests that reality is multilayered. "There is no sense in which, for example, subatomic particles—with their properties—are to be regarded as 'more real' than say, a bacterial cell, a living organism, or a human person."[12] Thus, different levels exist in reality without any one being more real than another. Newness—new emergents—arise out of lower levels and subsequently become the fodder for higher levels. However, this does not mean the possibility for newness ends at any one of these "lower" levels of reality or that there is movement toward some final point (à la Teilhard de Chardin). This linear movement would fall prey to the old postcolonial critique that societies left out of modernity are somehow static, primitive, and locked in anthropological and natural history museums. With multilevel emergence, rather, every level is open and evolving; levels can interact in different ways; and no level can be reduced to or subsumed by (which is really another way of saying reduced to) another. Both the scientific methods of reduction and a foundational understanding of theology have been guilty of cutting off natural-cultural life from its living source. Furthermore, both scientific and religious epistemologies have been guilty of ignoring Horkheimer and Adorno's caution in *Dialectic of the Enlightenment*. In a world where "reasonable" is equated with "reality"—whether one takes a theological, scientific, or combined approach to reality—myths/foundations are constructed to fit all of reality into the human confines of reason. As evolving realities tend to be much more than reason, and escape reason, the result is a reification of a part of reality and a mistaking it for the whole of reality. The effect of this process is alienation and violence toward any others who do not fit into the most reasonable construction; in other words, the effect is madness. I am not arguing that we should abandon reason for pleasure, but rather that we should recognize the limits of reason in capturing reality. Finally, acknowledging multilevel realities as, in one sense, having their own unique "perspective" may help guide our thinking out of the trap of ethical anthropocentrism, while acknowledging that our context as human beings (among other things) always already means we cannot avoid epistemological anthropocentrism.[13]

In what follows, I suggest that a viable epistemology, a living one, ought to be one that recognizes the limits of reason (which always ends up in foundations or in circular argumentation) and opts for a dialogical method of infinite regress. Before I describe what I mean by an epistemology of "infinite regress," I first offer a metaphor to help evoke a sense of the type of context from which my own thinking emerges. This context is very much shaped by the fluid reality I have been trying to evoke in the opening pages of this chapter. Here my attempt is to evoke a sense of how we might be able to better embody this type of fluidity as it relates to our understandings of knowledge and the process of meaning-making.[14]

## Evolving Habitats for a Viable Agnostic Theology

Nothing quite marks the living experience of the late twentieth and early twenty-first century as aptly as "contextuality": the notion of existing in located positions. Identity politics, rights movements, liberation struggles, the results of colonization and communicative and economic globalization lead to an experiential acknowledgment of ourselves as located. However, they also spur us toward the end of thinking about the "end of history." Once we see that we are many—of many stories, of many religions, of many identities, many truths, many values—*and* we see that we are, as the Apollo 8 image "earthrise" and the problem of global climate change suggest, of one planet, we are at a loss for words that make sense of it all. Linear narratives, histories, don't really capture the reality of our era. Rather, some form of reverberating contextuality comes to the fore. In other words, we are male, female, heterosexual, homosexual, bi, trans, queer, black, white, brown, Latino/a, American, Japanese, Kenyan, and more generally "of certain descents," yet we are also animal, biological, planetary, ecological—that is, inextricably bound within the planet Earth. "We" are not really a "we" without some sort of context, but the "text" that we are "with" is never stable, and the "we" is always shifting with the text between greater and lesser degrees of inclusivity. What should we do with this paradox between the need to claim particular identities (though evolving) and affirm our planetary emergence and bodily existence (some sense of common context)?

Perhaps we could think of the "time" that takes place in this new space (which is the hyphen of all the "post-" thinking) as reverberation rather than as linear or cyclical. From the present time we look toward the past, but only from our present context, which is also always defined by our hopes/dreams and visions of future becomings. Who is to say whether the voice is coming from the future or the past? If you have ever stood between two buildings or in a canyon and heard a noise that seems as if it is coming from in front of you but is really coming from behind, then you will understand what I mean by this experience. The present reverberates in much the same way between past and future, except the mathematics and physics of sound do not offer such precision in discernment when it comes to the realm of the present experience of past and future. Is what we interpret as history where we want to go? Is what we interpret as history rooted in where we have come from? Both versions of history, those relying on origins and ends, are placed into question.

In a reverberating present—marked by infinite regress in terms of origin, the context that gives rise and shape to the sound, and openness as the sound travels outward and cannot be contained—our epistemologies should be geared toward dialogue, toward the recognition of multiple perspectives. There is no solid past and future, but the becoming present decides how we interpret the past in accord with what visions we desire for the future. This is the bold reality of the hyphen in the "post-": without knowing where we are going we cannot project certainty into the past. This is the Deleuzian-Guattarian rhizomatic ontology that recognizes contextual ethical responses without justification in origins and ends. The "re-verb" of *re-verb-eration* implies the re-iterations or the ability to respond. Response-able. Responsible. Responsibility for the knowledge that we are always already "with" text, and taking responsibility for that text, is the place where our thinking begins. This is the type of habitat that is provided by the viable agnostic theology I articulate in the rest of this chapter.

## Thinking without Transcendence: Infinite Regress and Immanence

As Catherine Keller and John Dewey, among others, note, the quest for certainty using the "light" of reason has caused much damage.[15] In the face of an uncertain reality characterized by constant change and flow, many of us search for that equilibrium, that harmony, that I have argued above ends up as the search for a cold death. A prominent ancestor of the quest for certainty, Descartes, posits cogito/God as the block of this energy-thought flow. The "thinking thing" and God become transcendent foundations for a feedback loop in which the two are reinforced by each other. The cogito provides the epistemological foundation while God provides the ontological. For Descartes, it is impermissible that reality not be founded in something immutable; thus he posits cogito/God: "And although one idea can perhaps issue from another, nevertheless no infinite regress is permitted here; eventually some first idea must be reached whose cause is a sort of archetype that contains formally all the reality that is in the idea merely objectively."[16] Again, foundations, rather than infinite regress or circularity, are preferred, because foundations stave off the fear of "slippage" into some sort of chaos. I argue that both circularity and foundationalism cut epistemologies off from ontology: from living-flowing being-becoming. With circularity all that matters is individual coherence: the thought travels from point *a* to *b* and back to *a* again, without *taking* context. With foundationalism context is again denied for a source that transcends context. This relationship between foundationalism and circularity is displayed in Descartes's relationship between the self and God. The cogito is the first certain thing he comes to (in positing the evil genius), and from this he must, for the sake of certainty in the thinking-thing, further posit a Good God: God justifies self, self justifies God. Would this be described as foundational thought or circularity? There may be no difference in the end.

Instead of a transcendent principle that justifies foundational eternal claims to which all reality must eventually capitulate, why not speak of permeable horizons or borders? For Bruno Latour, the concept of "the collective" makes such a move. With the understanding of reality as an evolving "collective," these borders of knowing can never be completely

closed off from further questioning and investigation. "Its borders, by definition, cannot be the object of any stabilization, any naturalization, despite
the continual efforts of the great scientific narratives to unify what brings
us all together under the auspices of nature."[17] Rather, there is a process
of collecting, stabilizing, destabilizing, and collecting again, ad infinitum.
In this sense, infinite regress is not seen as a problem for epistemology,
but as that which keeps the collective process open to the other. In this
process, "a gradient is going to be established between the interior of the
collective and its exterior, which will gradually fill up with excluded entities, beings that the collectivity has decided to do without, for which it
has refused to take responsibility."[18] Outside externalities, then, transcend
the collective and make future challenges to it. It is in this process of collecting, breaking down, and collecting again, ad infinitum, that I think
knowledge begins to take place. Evolving epistemic communities with permeable horizons that shade off into unknowing prevents foundations and
circular arguments from reifying the living flow of thought (and life!) into
a cold death. Much like emergence and nonequilibrium thermodynamics,
Latour's collective and the infinite regress that it suggests opens up our
meaning-making processes and uncovers the evolving process of knowledge production. If we really trust that we are a part of the open, evolving
process of life, then why do we shelter our most creative-constructive edges
of thought, our meaning-making capacities, from this ebb and flow by the
use of foundations of certainty? A viable theology *of and for* the evolving
earth community, then, I argue, must be one that maintains mystery at its
open, permeable horizons. In a word, it must be agnostic.

## *A Viable Agnostic Theology* of and for *the Earth Community*

The claim that humans are radically in the world in no way negates the
uniqueness and wonder that marks the emergence of history and culture
from the ongoing process of continuing creation. "The gradual emergence
of human culture, of human activities and projects—has been as indis-
pensable as were the biological evolutionary advances that preceded our
appearance on planet Earth."[19] This "historicity" is what makes human
beings unique in the grand scheme of life.[20] It is made all the more awe-
some by the fact that it happens in such a vast universe. Historicity can
arguably be seen on this planet as accelerating the rate of change in continuous creation for better or worse. "The question is not whether humans
have induced change in ecosystems, but whether they have inordinately
accelerated or inhibited change and in ways that are deleterious, whether

to humans specifically or to the terrestrial life forms in general."[21] History/ culture is a "forcing" of biological evolution that emerges from biological evolution rather than something exceptional to the evolutionary process. Furthermore, human uniqueness from this bio-historical understanding of the human being does nothing to negate the uniqueness of all other life forms. From this position of bio-history, human beings (and here I am speaking particularly of Western human beings) can begin to speak theologically about the worlds in which we become.

As bio-historical creatures existing as part of and within the evolving, living world, theology moves away from metaphysics and foundational claims about God, at least from the "substance" metaphysics that characterizes much of the history of Western and Christian thought.[22] If we are not merely bios, nor merely history, then life cannot be reduced to either term. Reductionism and idealism aimed at forcing all of life to capitulate to "nature," "reason," or "revelation" no longer has a place in theological construction.[23] Rather, theology becomes conversational. This is what Kaufman sees as a move from "first order" or polemical theology to "third order" theology. Third order theology recognizes itself as imaginative construction and contends that "we must now take control (so far as possible) of our theological activity and attempt deliberately to construct our concepts and images of God and the world; and then we must seek to see human existence in terms of these symbolic constructions."[24] This also resituates truth from a transcendent space of foundations, to the evolving nature-cultures in which we live. "In this model, truth is never final or complete or unchanging: it develops and is transformed in unpredictable ways as the conversation proceeds."[25] Again, this model represents a viable agnostic theological and epistemological understanding of the world.

As response-able, bio-historical creatures, we can modify our symbolic structures to let more of life in rather than seal ourselves off from life in our concepts. These new reconstructions are not, of course, ex nihilo. Ex nihilo newness, as Keller has so insightfully pointed out, has been a major source of colonial violence.[26] Kaufman suggests, "It should not be supposed that the theologian creates the order into which he or she fits the multifarious features and dimensions of life simply *ex nihilo*. . . . it is always based on the prior human constructive activity which produced and shaped the culture."[27] Even those people who are considered radical revolutionaries cannot escape the bio-historical trajectories in which they exist. We can and do move and shape these trajectories, but we never move fully beyond them or into a "no-where" space outside of them. In this sense, Western theology is an integral part of the understanding of those who have grown up in or been influenced by Western culture. "Western theology should be seen first of all, thus, as a part of the language and traditions which have shaped Western experience."[28] This makes God-talk all the more relevant in a culture that segregates religion outside of the "public" realm.[29]

This understanding of theology as imaginative construction, as conversation, as the ongoing process of modifying living symbols in no way negates the reality of what we might call the "spiritual," "cultural," or "historical" side of life. This is often the critique waged against theological projects that fall within the trajectory of Feuerbach's insights—namely, that theology is thereby reduced to merely an individual projection and that therefore "God" is not "real." However, what our "bio-historical" existence means is that the "historical" side, the "cultural" side, and the "God" side *are* as real as the "bio" or "natural" sides of the equation. "To regard God and the world as constructs with which we bring order and personal meaning into experience does not involve downgrading them in comparison with directly perceivable objects. . . . indeed, we could not live and act and think without some such ordering principles or images."[30] Our concepts, our theologies, and our symbols are as much a part of our *real* life as those things we directly perceive. Indeed, this is one of the implications of bringing God and the world back together in an ongoing process of continuous creation. Despite some of Feuerbach's own negative impressions about theology—that concepts of God are almost always unhealthy egoistic projections—many have suggested that Feuerbach can actually be taken in a different direction (and himself was moving in that direction), seeing theological projections as part of what it means to be human.[31] "In other words, the question is whether we can take the imaginative nature of God's origin not as an indicator of its non-reality, but rather as a sign of the poetic nature of God language and hence of the symbolic nature of God's reality."[32] In many ways this is what all "post theologies are about. This is an attempt to *make real* the idea that whatever it means to be human includes meaning-making and that these effects of meaning-making—value, imaginations, hopes, dreams, theologies—are indeed just as real as chemicals, neurons, genes, atoms, and so forth. What would differentiate this type of theological projection from the type that Feuerbach critiques would be precisely the agnostic, post-foundational character of the projections. These imaginings are precisely not locked into the real/not-real structure of Feuerbach's own modern mind but, rather, comprise a unified reality. "The importance of Feuerbach's thesis that the imagination is the key to understanding how and why human beings are religious lies beyond the horizon of his own narrowly positivist epistemology and

stolidly modernist temperament."[33] Though Feuerbach's modified understanding of religious thinking/theology as projection/imagination may
offer an evolutionary-emergent way of understanding religion as part of
the natural-cultural world, it no longer need signify an evolutionary su-
percessionist understanding of the development of religious ideas. That is,
the move from "primitive" and "animistic" to "transcendent" and "mono-
theistic" ways of thinking about religion are no longer a part of a historical progression toward those in power now (i.e., Western Christians) but,
rather, can be interpreted as clues to how various natural-cultural worlds
have been constructed throughout time, for better and worse. These past
ways of co-constructing meaningful worlds *reverberate* in the present and
enable us to imagine together a future. These religious imaginings do not
become justification for a way of being in the world, but become part of
the contemporary effort to imagine again how we want to become. There
is no pre-given end; projections and imaginings do not close off reality, but
remain open toward a radically unknown future.

I cannot stress enough the fact that this type of agnosticism is precisely
not atheism. Robust theisms of the systematic or fundamentalist stripes and
robust atheisms of the Richard Dawkins type both say way too much about
what we do not know—namely, the places beyond our sensory capacities
are filled with either the foundation of God or Nothing. This is precisely
an illustration of the problem of using foundations to take us outside of the
rest of the evolving, open-ended natural-cultural worlds. Neither side will
ever listen to the other; instead, both will try to make the other conform
to its way of thinking. What other options are there from within a closed
system of thinking? How is any recognition of the truth claims of an other
*really* possible when meaning-making practices are seen as given, whether
that means *given* in the form of nothing or ultimate meaning?

What I suggest is that a viable agnostic theological method does not
take meaning away from the world, but has implications for the scope and
scale of our meaning-making projects. In foundational understandings of
reality, human meaning must be narrated across the span of the universe
from origin to end. In circular understandings of time, meaning is all-
inclusive; everything has its proper place and is accounted for. The re-
verb sense of time implied in this theological method is more akin to what
Keller describes as fractal-like, or what Deleuze and Guattari describe as
rhizomatic. In Keller's understanding of creation as pluri-singular, time is
fractal- or spiral-like; it is not mere cyclical repetition, nor a linear movement that cuts creation into isolated moments, but spiraling recapitulation.
Just as Deleuze and Guattari's notion of rhizomatic thought works against

the tendency to think in linear terms, so Keller's understanding of time could be described as rhizomatic. In this understanding of time, the scandal of particularity becomes the norm of particularity: "Any event, every space-time of the capacious process of creation, might become readable as a unique, holy, temporary embodiment of the infinite."[34] Thus, the universal *is* the particular—all of the particularities in the becoming creation are universal. In other words, there is no "outside space." Following Derrida, Keller notes, "Nothing left outside: this might entail simple atheism and the collapse of meaning along with the space of transcendence. Or it might mean that the divine can no longer be situated in a standard theistic beyond and therefore that it cannot offer the false sense of security encoded in the classical construct of the changeless transcendence, with its omnipotence that could unilaterally intervene to reward, punish, or rescue."[35]

My hope for this theological method is something like an interdisciplinary conversation process in which we make explicit our implicit practices of meaning-making and begin to take responsibility for them. In this way, we can begin to take note of those things that are left out of or harmed by our systems of meaning. This is akin to the pragmatist vision of a John Dewey or a Richard Rorty. Furthermore, it turns meaning-making processes into a kind of participatory democracy rather than an elitist, trickle-down type of activity.[36] Just as science cannot dictate "nature," so here religious leaders and theologians cannot dictate "meaning." Nor can any "private" belief dictate religious meaning. As Rorty notes, beliefs are never private: "There is no way in which the religious person can claim a right to believe as part of an overall right to privacy. For believing is inherently a public project: all us language-users are in it together."[37] Rather, both scientists and theologians help us to tussle with what these tropes of "religion" and "nature" might be. Philosophers Deleuze and Guattari also capture the type of interdisciplinary knowledge production I am hoping for in their understanding of "nomadic thinking."[38] Though we travel in time and space on textured grounds, our thoughts never land on terra firma, because the ground, as Keller notes, is precisely not foundation.[39] It is always shifting and moving. If we do not recognize this located-ness of thought, the towers of language and symbols we decide to stay in will mistake shifting grounds for foundation and then begin to remake and force all of reality into the image of itself.[40] This is precisely the reason that a *viable* theology must be agnostic.

Any new symbolic constructs, including theological, will be located in a bio-historical context and thus will have certain blind spots and the need of continuous transformation. This does not mean we are *reduced* to the specific bio-historical context in which we find ourselves at any given moment (another form of foundational thinking), but that it is the only space from which we can see at all. There are temporal horizons between which the present exists: we can "see" into the past and "look" toward the future, but never beyond to any foundational point.[41] However, we can never seal off the present from the ongoing process of continuous creation. In this sense, the discipline of theology has to be thought of in terms of the changing environments around us: as a piece of "environmental history" rather than as somehow merely imposing thought onto matter.

In terms of a theological epistemology, this dialogical thought involves recognition of the location of rational thinking within the context of evolving epistemic communities, and as a human capacity rather than some context-free space in which we participate. J. Wentzel van Huyssteen calls this rational capacity the human ability to "make sense" of and "cope" with the world.[42] Kaufman refers to this capacity, which extends even to theological reasoning, as the ability to critically converse.[43] Theological reasoning as conversation implies that there are multiple dialogue partners in the process of theological knowledge. Etymologically, it also related to the willingness to be changed (to live *with*): conversion. This openness to the other is the place from which dialogical theological exploration can begin. Rather than coming to the conversation in a monological way, with foundations that cannot be moved but must be accepted, we can come to the conversation with openness to the possibility of being changed by the other, which always already happens to begin with (whether recognized or not).[44] However, openness to the possibility that theological exploration might convert you is not the sole purpose of theological knowing. Rather, theology has always also been about "making sense" (or making meaning and value) of the world, and this world, from a human perspective, is a natural-cultural world.

Given that we exist in relationship to human and earth others, theological reflection must also include our whole experience and the experience of others. One objective of this viable, agnostic project is to expand theological and meaning-making practices beyond the individual, group, or humanity in general, to all of creation, with the recognition that we cannot get beyond epistemic anthropocentrism (as Val Plumwood makes clear) and that we are cocreators (with all other life) of the worlds in which we live. For Christians, theology will always be about the experience *of* at least Christian humans, but that does not mean it is only *for* Christians or even humans. This is precisely *not* a call for making the rest of the world conform to Christianity, but rather a call for those working from within

Christian meaning-making structures to take note and responsibility for
the ways their meaning-making practices affect the rest of humanity and
the rest of the natural world. It is precisely contextual in that it argues
against both: (a) universalizing/colonizing the world with a particular idea,
thought, or imagining; and (b) relativizing one's own way of thinking and
imagining to the extent that he or she does not even take responsibility for
how it affects others.

I am extending theology as projection beyond what is perhaps a misin-
terpretation of Feuerbach's project as anthropological and somewhat individualistic version of projection. For Feuerbach, "God as such, the one
universal God, from whom the bodily, sensuous attributes of the many gods
have been removed, does not transcend the genus *homo*; he is only the most
objectified and personified generic concept of mankind."[45] The reason for
extending theological projections beyond the personal is that the worlds
in which we live are much more than the personal. Our theologies are *for*
the evolving nature-cultures in which we live, though they, like all systems
of thought, are epistemically located in specific bio-historical human experiences. Like all other symbols, theological symbols shape the ways in
which our nature-cultures evolve. Though they are intangible, theological
constructions actively shape our worlds. Theology is projection, but projection that is not merely individualistic; it is a continuous projection of
evolving communities' experiences of the world that *matters*.

Some may still argue that the method I am articulating is atheistic, just
as some interpreted Feuerbach's claims about theology likewise. However,
I contend that atheism and theism are really two sides of the same coin, as
are reductionism and holism, or relativism and universalism. The former
of each pair reduces reality to the self, at least in some versions, while the
latter makes the self in the know about the whole (for example, we are "in
the image" of God and through theological inner-course become *like* the
omni-God we imagine, ordering all of life according to this knowledge).
In both cases, we are taken out of the depths of ongoing creation; both in
a way reduce all to self. As Mark Taylor suggests, "If the master is God
and the slave man, then man's murder of God is an act of self-deification.
. . . The death of the sovereign God now appears to be the birth of the
sovereign self."[46]

In a theology that doesn't stop at projected foundations, everything
comes from something. Thoughts and knowledge come with textures that
can be reflected upon, interpreted, and reconstructed as something new
(but not from nowhere) for the present bio-historical context. "My" theology is not then my own projection: it comes from centuries of "others,"

both human and non-, as the nonhuman "other" always already shapes and enters into our theological reflection. I am response-able (and in many cases responsible) in this reconstruction project, but not fully in charge or fully in the know. Because the process of knowing is always already from within specific bio-historical contexts, and our knowledge systems are always interpreted anew. We cannot know how our reconstructions will affect all of the life around us, but we can be open to the process of continuous creation, open to being changed by "others" in conversation, and open to how our theologies affect earthly others.

At the edges of our theological projects, then, we must admit, and learn to love as part of our meaning-making bio-history, the darkness, the unknown, the mystery. It is from these dark matters, these unknowns, and these mysteries that the creative-destructive process of life becomes re-energized, renewed. Once these dark spaces are filled with the projected light of certainty and the living stream of life is "damned" up, equilibrium in nature and culture will lead eventually to cold death.[47] Therefore, I argue, a nonequilibrium, and thus viable, theological understanding of the world should also be agnostic. We need to relearn the virtue of unknowing to get beyond our quest for certainties, or even as Talking Heads suggests in an album title, and as philosopher John Llewellyn might agree, we need to "stop making sense."[48] As we begin to recognize that our meaning-making practices find their groundings in evolving contexts that shade off into mysteries, and our sense-making loses its control of the edges of our understandings, perhaps we will begin to open up toward the evolving, planetary contexts in which we exist and begin co-constructing with many earth-others a more inclusive, participatory planetary democracy.

# PART II

# Engaging Ecology and Culture

CHAPTER 6

# Ecological Civilizations: Obstacles to, and Prospects for, Religiously Informed Sustainability Movements in a Post-American World

*Jay McDaniel*

> Oil production is already leveling in China, Mexico, and India, but without the alternatives already in place. Water supplies are diminishing in China and the Middle East. This leads to the possibility of inter-state conflicts over resources, with regimes deeming it necessary to battle others to maintain stability in their own nations.
>
> *GLOBAL TRENDS, 2025*

We live in an age of declining energy reserves, and it is likely that by 2025 nations will be battling other nations in order to secure scarce resources. At least that is what a recent report from the National Intelligence Council proposes. Optimistically minded theologians might counter that if people lived with compassion and a willingness to share resources, the resources might not be so scarce. Finite resources are real, they say, but scarcity lies in the eyes of its beholders. Theologians with more suspicious leanings might add that the rhetoric of scarcity is but a tool used by powerful people to hold the poor in bondage and engage in profitable warfare. There is wisdom in both of these perspectives.

Nevertheless, our finite energy reserves do have their limits, and our compassion may have limits, too. Whether we adopt a hermeneutics of trust or suspicion, we cannot avoid the fact that if violence is to be averted and life is to flourish, nations must evolve into what the Chinese call "ecological civilizations." These civilizations simultaneously embody two kinds of harmony: harmony with the earth and harmony among people. Harmony with the earth entails living within the limits of the earth to absorb pollution and supply resources; harmony among people entails living in ways that are socially just and spiritually satisfying with no one left behind. These two kinds of harmony are the building blocks of a holistic sustainability that includes, but also goes beyond, mere environmentalism.

If ecological civilizations are to evolve, it is possible that life-affirming religion can play a constructive role in this development. By "life-affirming religion" I mean religious traditions that elicit in their advocates and practitioners an attitude of respect and care for the community of life, humans and nonhumans included. Not all religion is life-affirming, and most of the historical religions contain elements that are life-negating as well as life-affirming. Nevertheless, as the work of Yale University's Forum on Religion and Ecology has made clear, every religious tradition contains seeds of life-affirmation, and these seeds are worth cultivating.[1]

Moreover, religions are evolving over time, and even if they lack seeds for sustainability now, they can develop them in the future, since no religion is fully defined by its past. These seeds of life-affirmation can help people realize that energy is not reducible to the kind of energy found in oil and water but also includes the powers of empathy and imagination whose supplies, while not unlimited, are more abundant than imagined. Perhaps the theistic seeds can help people realize that there is a wellspring of divine compassion at work in the universe—call it the "Deep Listening"[2]—whose own supply of energy supplies sustenance for daily living and hope in time of need.

I speak of the divine compassion as a Deep Listening in order to accent the fact that there is a certain kind of energy—both human and divine— that is powerful in a receptive way, not a causal way. When we are in the presence of people with open minds and receptive hearts we sense this receptive power. What makes them powerful is not that they can cause other things to happen, but rather that they can include so much within the space of their heart-minds. Perhaps theists can help people imagine God as a spacious receptacle or open space whose life includes the hills and rivers, trees and stars, and in whose image humans are made. This way of imagining God may not seem theistic in a customary sense. It does not envision God as entirely separate from the world. It is instead *panentheistic*. It says that everything is included within the divine life even as God is more than everything added together. Perhaps one of the better hopes for theistic religion is that it can encourage panentheistic understandings of God. Such understandings might help them understand that an impoverishment of the earth's biological diversity is an impoverishment of the divine life, too. Of course the seeds of panentheism are already found in many of the world religions. The key is to water the soil.

At least this is how things look from the vantage point of the remainder of this essay. My aim is to amplify some of the comments made above with help from *process thought* or *constructive postmodernism*. I hope that some of the topics introduced, albeit all too briefly and cursorily, might provide food for further reflection.

## Constructive Postmodernism

In the West *process thought* usually refers to ways of thinking that are indebted to the philosophy of Alfred North Whitehead or Charles Hartshorne. The term *process thought* is preferable to *process philosophy* or *process theology*, because a growing number of process thinkers do not teach in departments of philosophy or religious studies, but are instead in departments of education, politics, economics, or literature. Indeed, and unfortunately, there are very few departments of philosophy in the United States that offer opportunities for a thorough study of Whitehead or Hartshorne. However, in other parts of the world, especially in Europe and Asia, opportunities for studying Whitehead are increasing. Four of the most recent International Whitehead Conferences took place in Beijing (2002), Seoul (2004), Austria (2006), and Bangalore (2009). Many process thinkers believe that a leading edge of process thought—perhaps even *the* leading edge—lies in China. This belief reflects the fact that for many Chinese intellectuals the relevance of process thought lies not only in its theoretical depth but also in its practical applications in the face of modernity.

In China process thought is also called *constructive postmodernism*. The phrase originated with David Ray Griffin in the West but has been adopted and used more widely by Chinese than by Western process thinkers. Often Chinese process thinkers are asked why they use the word *postmodern* to name their Whiteheadian way of thinking, given that the phrase naturally refers to European deconstructionists such as Foucault, Derrida, Lacan, Kristeva, Levinas, Lyotard, and Deleuze. The answer is partly an accident of history. The leading advocate of Whiteheadian postmodernism in China, Zhihe Wang, wrote an important book on European postmodernism that introduced Chinese scholars to the movement. Wang concluded his book with an appreciation of certain aspects of deconstructionism, but also with the suggestion that deconstructionism is inadequate for contemporary Chinese intellectuals because it offers little in the way of a more positive and constructive approach to the most serious problems that China faces today, including environmental problems. He found himself looking for something more constructive and turned to Whitehead for that option, bringing Griffin's phrase with him. But this is not the whole story. In the context of contemporary China, the word *postmodernism* can have a very positive connotation, insofar as it suggests an alternative to the reigning problem in China today, which is how to "modernize" without "Westernizing."

Constructive postmodernists in China share certain common themes with European postmodernists. Both groups welcome diversity and difference; both reject substance-oriented views of the self; both are critical of meta-narratives, including their own, when they function to legitimate unjust forms of social control; both see the world in terms of events and relations; both recognize that language cannot itself be reduced to univocal meanings. Moreover, as Catherine Keller makes clear in her work, and as Roland Faber likewise makes eminently clear, it is possible to combine the insights and sensibilities of Whiteheadian and European postmodernism in ways that enrich both and in ways that are deeply *constructive*. This suggests that the dichotomy between deconstructive and constructive postmodernism is not as sharp as some might suspect. As Keller shows, a deconstructive approach can be used to offer constructive approaches to environmental problems.

Nevertheless, Whiteheadian postmodernists in China think of themselves as "constructive" rather than "deconstructive." Whiteheadian postmodernists are committed (1) to the development of integrative worldviews that link science and spirituality, concern for human beings with respect for the earth; (2) to the encouragement of forms of spirituality and wisdom that draw upon the best of inherited traditions and the best of modernity; and (3) to the cultivation of social practices that can help people of different nations—China and the United States, for example—overcome environmental problems and build sustainable communities in rural and urban settings. Thus, constructive postmodernists are postmodern not in the sense of being anti-modern, but rather in the sense of trying to build upon the best aspects of modernity and tradition, thus creating new ways of thinking and acting in the world. It is in the sense of this kind of Chinese postmodernism, then, that I use the word in what follows.

### The Post-American World

A 2008 report from the National Information Council (NIC) called *Global Trends, 2025: A Transformed World* tells us that the age of American hegemony is in decline and that citizens throughout the world will soon be living in a post-American world. Developed by a team of researchers whose task is to guide American foreign policy, the report predicts that if present trends continue, by 2025 the United States will be one power among many in a multipolar world. Admittedly, the United States will still have much economic, political, and military power. But its power will be challenged, balanced, enriched, or thwarted by other national powers: China, India, Iran, Russia, and Brazil. If all the world is a stage and the men and women are its players, then many nations will be at center stage.

For many Americans this report is good news, not bad news. For process thinkers in the United States it invites the emergence of a new kind of patriotism—a postmodern patriotism—in which Americans take pride in some of the better features of our nation and its many cultures, but in which we also take delight in the beauty of other nations and cultures. It simultaneously encourages the emergence of a new kind of foreign policy— a postmodern foreign policy—in which we seek win-win situations with other nations and celebrate opportunities for partnership, not domination. And it encourages a new approach to globalization—a postmodern approach—in which we distinguish between healthy globalization and unhealthy globalization, embracing the former while rejecting the latter.

Of course there are many kinds of globalization: economic, political, cultural, educational, electronic, and spiritual. Each kind can be healthy or unhealthy. *Unhealthy* economic globalization is a destructive "integration" of global financial markets resulting in suffering in many parts of the world. Problems in the United States banking industry, due to greed and an absence of regulation, can lead to 20 million unemployed Chinese migrant workers. From a process perspective, *healthy* economic globalization must begin with the development of stable national economies in which nations achieve relative self-reliance on basic needs such as food and energy and then trade among less essential goods.

One of the most discouraging proposals of the NIC report is that in the year 2025 some countries may be fighting over food, water, energy, and other scarce resources. If the world is a stage, then the stage is the earth itself. The actors will include global warming, polluted waters, depleted topsoil, shrinking forests, expanding deserts, and, by then, the political instabilities that arise from them. Demand for food will rise by 50 percent by 2050, says the report, with more than thirty-six countries finding themselves with inadequate cropland, fresh water, and food.

Global warming will intensify the problem, leaving nations in sub-Saharan Africa most vulnerable. Demand for energy will also be a problem.

Oil production is already leveling in China, Mexico, and India, but without the alternatives already in place. Water supplies are diminishing in China and the Middle East. This leads to the possibility of interstate conflicts over resources, with regimes deeming it necessary to battle others to maintain stability in their own nations.[3] These conflicts can be averted only if nations learn to live within the limits of the earth to absorb pollution and supply resources.

### Ecological Civilizations and Sustainable Communities

How, then, might the many nations of the world step forward and avoid such disasters? Here it seems to me that the Chinese government offers the needed vision. Their vision is that the many nations of the world evolve into "ecological civilizations" that combine two forms of harmony: harmony with the earth and harmony among people. The idea was officially proposed by Hu Jintao at the 17th National Congress of the Chinese Communist Party in 2007, and it has now become part of Chinese public policy.

President Hu did not offer this vision in a vacuum. He recognized that environmental problems are severe in China and that the nation is far from being an ecological civilization. The report from the *China Daily* was brutally honest:

> This concept [of ecological civilization] is proposed at a time when
> 62 percent of the country's major rivers have been seriously polluted,
> 90 percent of waterways flowing through urban areas are contaminated,
> more than 300 million residents are yet to have clean water to drink,
> and quite a number of localities fail to fulfill the required quotas for
> pollutant emission reduction and energy saving.[4]

Thus, for the Chinese government, the concept of an "ecological civilization" is an ideal to be approximated. It is meant to be "a future-oriented guiding principle based on the perception of the extremely high price we have paid for our economic miracle."

This future-oriented principle has relevance to the United States, too. Of course, most Chinese and Americans do not live from this kind of ideal. Arguably, the dominant dream of both nations today is to make money. Nevertheless, there are deeper hopes within the minds and hearts of most Americans and Chinese. In China there is the hope of becoming a socialist society in which people share in one another's destinies, and in the United States there is the hope of becoming what Martin Luther King Jr. called a beloved community. In both instances the hope is for what we might call sustainable communities.

From a process perspective, the building blocks of ecological civilizations are indeed sustainable communities. *Sustainable communities* can be defined in different ways. I find it helpful to distinguish between "environmental sustainability" and "inclusive sustainability." The latter is what is recommended by process thinkers, but it includes environmental sustainability. *Environmental sustainability* occurs when a community lives within the limits of the earth to absorb pollution and supply resources at rates that do not exceed carrying capacities, and the community makes space for other forms of life by protecting wilderness and endangered species. They dwell in harmony with the earth.

*Inclusive sustainability* occurs when the conditions of environmental sustainability are approximated and when, in addition, human beings enjoy meaningful interactions with one another with no one left behind. Meaningful interactions include social justice, both economic and political. But they also include opportunities for rest and relaxation, friendship and leisure, and for the enjoyment of beauty. Communities are inclusively sustainable if they are creative, compassionate, equitable, participatory, pluralistic, ecologically responsible, and spiritually satisfying—with no one left behind. The communities at issue can be villages, towns, cities, or provinces; households, workplaces, schoolrooms, universities, or churches. Whatever their scale, they are "sustainable" in two senses. Their citizens live within the limits of the earth to absorb pollution and supply resources, and they will provide sustenance for human life.

Objectively, an approximation of sustainable communities requires the adoption of sustainable policies and practices on the part of government, business, and education. Here economists, politicians, businesspeople, bankers, agriculturalists, architects, and urban planners have special roles to play. They must help create conditions where people have healthy food to eat, satisfying employment, good health care, and social security.

Subjectively, the approximation will require the emergence of a certain form of consciousness—a sense of respect and care for the community of life—on the part of people who live within them. Here philosophers, theologians, artists, and educators have a special role to play. They must help develop ideas that inspire people to want to create sustainable communities in the first place. Of course bankers can be philosophers, too. And artists can be politicians. There is no need to compartmentalize people into one and only one vocation.

## The Role of Religion in Helping Bring about Sustainability

Thus the question is: can the religions of the world help elicit attitudes that help bring about sustainable communities? The answer depends on four factors: (1) the degree to which people who are "religious" are actually influenced by the "religions" with which they identify; (2) whether the societies in which they live allow "religions" to play a role in decision making; (3) the question of whether in being "religious" they draw upon the more life-affirming themes within their traditions; and (4) the relation of religious self-identification with other forms of self-identification, which may be equally if not more powerful and which may offset more constructive religious tendencies. In many countries the psychological power of nationalism overrides that of religion. A person can be nominally Buddhist but deeply Chinese; nominally Muslim but deeply Palestinian; nominally Christian but deeply American. And it is possible that "being Chinese," or "being Russian," or "being Capitalist" may counter whatever it might mean to "be Buddhist," or "be Muslim," or "be Christian."

This is not to say that being "deeply religious" is always a good thing. One of the problems with religions is that people who adhere to them establish their identities by objectifying and then demeaning others. This is especially true for people who "belong" to religions that stress exclusive and totalizing allegiance. If they are overly fervent, their religions become their gods, and they lose their capacities for self-criticism and a respect for others. In Christian circles today there is a great emphasis among evangelicals on being "true to our Christian identity" or "true to our roots." When this occurs Christianity becomes a fence by which other people are shut out. If a sustainable world is one where people are truly hospitable to one another, then the world may need less religion. Or at least less fervent forms of religious self-identification. The world may need "weak religion," not "strong religion."

In this regard we do well to consider the writing of one Buddhist writer from Thailand, Sulak Sivaraksa. He encourages Buddhists to become Buddhists with a small *b*. By this he means that Buddhists ought to approach Buddhism as a means to an end, not an end in itself. The end is a reduction of suffering among living beings and a promotion of healthy community, not the promotion of Buddhism. Of course Buddhists are relatively small in number. They constitute about 6 percent of the world's population. On the other hand, Christians constitute about 30 percent of the world's population and Muslims 20 percent. It is arguable that if Christians and

Muslims are to serve the interests of a sustainable world with help from the better ideals of their religions, they must become less preoccupied with their own religious identities. They must become Christians with a small *c* and Muslims with a small *m*.

All of these considerations play a role in determining whether "religion" can help bring about a more sustainable world. The straightforward answer is: it depends. Even if a tradition carries helpful ideas, the potential value of a religion for a sustainable world depends not only on the content of the religions themselves but also on the place of religion within people's self-identities, the manner in which they adhere to the religions to which they belong, and the extent to which a society can make decisions based on those values.

The matter is complicated by the fact that no one is quite sure what religion is and how it is different from culture. There was once a time when intellectuals in the West thought that religion always deals with God or the Gods. But if we define religion in this way, then the category of "religion" does not include many kinds of Buddhism, Taoism, and Confucianism. Nor does it include what is arguably the dominant religion of our time: consumerism. It is fine to exclude these traditions from what we call religion. But then the very category of religion becomes less interesting to intellectuals who are concerned with the formation of attitudes conducive to a sustainable world. Culture seems more relevant than religion.

But here the problems surface anew. We can overestimate the role of culture in shaping attitudes of people in the world, and we can also underestimate it. It is at this point, it seems to me, that the philosophy of Whitehead offers a helpful category that includes some of what people mean by "religion" and "culture." It is his idea of *subjective aim*. The subjective aims of a society consist of the ideals around which its people organize their private and public lives. In Whitehead's philosophy an individual can have subjective aims, and so can societies.

One Whiteheadian thinker, John Cobb, proposes that in our time many societies in the world are torn between two dreams. One is the dream of economism, the hope that most of the important problems in society can be solved by means of a perpetually growing economy. Some then add that the market must be regulated by government, and some say it ought to reign supreme. In China, for example, people who subscribe to the dream of economism generally appreciate the role that government can play in economic enterprises. They exemplify market socialism with Chinese characteristics or what others call state capitalism. In the United

States, by contrast, people who subscribe to the dream of economism have often emphasized the importance of unregulated markets.

The other is the dream of ecology. This is the hope that many important problems can be solved, and that people can live happily, if they learn to live with respect and care for one another and the larger community of life, and if they then adopt public policies that are aimed not at ever-increasing economic growth, but at the well-being of life, both human and ecological. People who subscribe to the dream of ecology do not eschew growth. But they distinguish between "healthy growth" and "unhealthy growth" and then add that any kind of growth must be in service to community.

The dream of ecology, of course, is the dream of inclusive sustainability as defined above. The word *ecology* is used as a metaphor for mutually enhancing relations between humans and the natural world and also for mutually enhancing relations among humans themselves. In this view, justice is a form of ecology, and so is environmental protection. The central question of our time is whether "religions" and "cultural traditions" can encourage attitudes and practices that are conducive to the ecological aim.

## The Vocation of the Ecotheologian

The vocation of the ecotheologian in our time is to encourage "religions" and "cultural traditions" to foster these attitudes and practices. Here I use the word *ecotheologian* in a very general sense. I refer to an intellectual, working within or outside the parameters of a university, and within or outside the context of a recognized discipline such as "religion" or "philosophy," whose aim is to encourage respect and care for the community of life.

Ecotheologians working within the various world religions generally adopt one or both of two approaches: historical and constructive. Historical ecotheology is directed toward examining inherited religious and cultural traditions in order to critique attitudes and practices that lend themselves to a domination over other people and the earth; and to identify and affirm attitudes and practices that lend themselves to respect and care for the community of life. An excellent example of this approach is the foremother of contemporary Christian ecofeminism, Rosemary Ruether. Her pioneering work *Gaia and God* examines what she calls the "usable" and "unusable" past of Christian intellectual history. She concludes her work by proposing that there are two traditions within historical Christianity that are usable, given the aim of helping encourage the development of sustainable communities: (1) the covenantal traditions that speak of a transcendent God who establishes a covenant with human beings and also with the earth, and who mandates humans to be good stewards of the earth; and (2) the sacramental traditions that find something holy or sacred in the web of life itself. The covenantal are Protestant in orientation, and the sacramental are Catholic.

The more constructive kinds of ecotheology are not content simply to examine the inherited traditions, but they also seek to construct new ideas that help advance the ongoing history of a tradition. An excellent example of this kind of ecotheologian, also in the ecofeminist lineage, is Catherine Keller. Using insights from process thought, but also drawing deeply from biblical and historical traditions, Keller offers Christians and others fresh ways of recognizing fresh opportunities for living with respect and care, while at the same time avoiding the stale and sometimes oppressive rhetoric of predictable Christian thinking. She is able to do this, in part, because she has internalized the spirit of the deconstructive movement discussed earlier. While her sensibilities lean more toward the sacramental than the covenantal, and oftentimes toward the mystical sides of sacramentalism, Keller embodies two additional strands of historical Christian traditions: the prophetic and eschatological strands. The prophetic strand feels free to "speak truth to power," including the power of oppressive rhetoric, and the eschatological strand feels free to "sing a new song," even if it has no exact parallels in the past.

The role of God in the ecotheologies of Ruether and Keller can seem unclear to those outside and within the Christian tradition, because neither uses the word God with ease. Both authors are familiar with the ways in which the very word has functioned in oppressive ways in human history and with the fact that it can sometimes divert attention away from this world toward a transcendent authority. In this respect both ecotheologies point Christians toward what might call the *horizontal sacred.*

The horizontal sacred is this very world, understood as a locus of intrinsic value. It is best understood in comparison with the vertical sacred, which is the subject matter of much traditional theology in the West. By "the vertical sacred" I mean God, understood as a higher and sacred power of one sort or another. As a process theologian, I myself believe in such a power, and I say more about this in the last section of this essay.

Ecotheologians from the many world religions do not necessarily center their thought in terms of such a power. Instead they orient themselves toward the value of this world for its own sake, God or no God. To be sure, many humanists orient themselves toward this value, but their focus is primarily on human beings. They say they find something sacred in the face of the other, in the value of the stranger, in the worth of each individual human being. Ecotheologians add that they find something sacred—or at least worthy of deep respect—in the more than human world as well: the hills and rivers, the plants and animals, the trees and stars. In this sense ecotheologians go beyond humanism, without leaving humanism behind.

In Western circles this affirmation of intrinsic value is sometimes grounded in God. Theists speak of God as the source of intrinsic value or, in some cases, as intrinsic value itself. Those with sacramental sensitivities will say that when we gaze into the eyes of another person or animal we are seeing the very face of God in the face of the other. A unique feature of process thought, though, is that it recognizes that all living beings have value worthy of respect. From a process perspective, an adequate ecological ethic will begin with a recognition of three kinds of value: the value that living beings have in and for themselves, the value that they have for one another, and the value that they have for the inclusive whole of the universe, namely God. God is not the source of value, but rather the catalyst for value and the receptacle for value.

In the final section of this essay I want to say more about the process understanding of God in relation to energy. Process thought offers what is often called a panentheistic approach to God. Of course panentheism is different from pantheism. *Pantheism* is the view that God is identical with everything in the universe; *panentheism* is the view that everything in the universe is part of God in some way, even as God is more than everything added together. The critical question is: what is this moreness? Many Western monotheists think of the moreness of God on the analogy of the subject of a sentence to which predicates are added. God is the subject and God's actions in the world, and the world's actions in God are then predicates. The predicates could be eliminated and God would remain God. A unique feature of process thought, though, is that subject-predicate ontologies are abandoned and a more relational point of view is offered in their place. This has many implications for understanding the relationship between God and energy, which in turn has implications for a process approach to sustainability. I turn, then, to a discussion of energy.

## *Many Kinds of Energy*

There are many different kinds of energy: geological, emotional, intellectual, moral, imaginative, aesthetic, and divine. Considered in itself energy is neither good nor evil, but it can be embodied in healthy or unhealthy ways. Cruelty and violence have a palpable kind of energy, but so do tenderness and care. Most people assume that divine energy is trustworthy energy and that openness to divine energy provides a kind of guidance for how to approach the other kinds.

The various cultures of the world have generally been sensitive to these many forms of energy. In the New Testament the Holy Spirit has a kind of power and is a kind of energy. In Indian Tantric traditions sexuality has power and thus one kind of energy. In Africa departed ancestors have power and thus a kind of energy. In China mountains and rivers have power and thus are energy. In Islamic societies the sounds of the Qur'an have power and thus are energy. It can seem odd that in the modern West we have sometimes limited energy to what can be understood by modern science and what can be used for commercial purposes. Science focuses on forms of energy that can be subjected to experimentation and interpreted mathematically; and commerce focuses on forms of energy that can be "held in reserve" and then used for industrial purposes. From a Whiteheadian perspective, these are indeed important kinds of energy. But they are not the only kinds. One can easily imagine an essay written by a Whiteheadian thinker called "The Liberation of Energy." Its aim would be to liberate energy—or at least the concept of energy—from its reduction to two forms: quantifiable and commercial.

Of course this raises the question: what *is* energy? There is no single definition of energy upon which all will agree. In a Whiteheadian context one way to define energy is to say that it is power. More specifically it is the power of an entity to influence others, and it is also the power of an entity to receive influences from others. Let us call the first kind of power *causal power* and the second *receptive power*. I must quickly add that the word *entity* can be very misleading. Ordinarily the word suggests something substantial and solid that endures over time, like a billiard ball. Whitehead does not use the word *entity* in this way. He believes that the truly real entities in our universe are happenings or events, rather than substantial and solid objects, and that solid objects are actually composed of such events, as are we ourselves. Thus he offers what some call an event-ontology as opposed to a substance-ontology. Scholars of Chinese philosophy—and Chinese process thinkers themselves—suggest that an event-cosmology is more typical of Chinese than Western points of view. This may be one reason that Whitehead himself proposed that in many ways his "process philosophy" was more Asian than Western in tone. In any case, for Whitehead the entities of our universe are events, and the most immediate example we have of such events are our own experiences, as lived from the inside. Writing these words is an event in my life, and reading them is an event in your life. These events are the "entities" that compose your life, and both exemplify the two kinds of power named above. Whitehead speaks of the power to influence something else as *causal efficacy* and the power to receive influences as *the act of prehending.*

The act of prehending is an act of feeling the presence of something else and thus taking it into account within the immediacy of experience. Whitehead uses the word *prehending* rather than *apprehending*, because he thinks that the act of prehending can be non-conscious as well as conscious. He further believes that in any given act of human experience these two forms of power go together. When we are listening to another person, we are receiving that person into our experience (receptive power), and through our listening the person is influencing our immediate state of our mind (causal power). If we were Chinese we might speak of this twofold dynamic—the giving of influence and the receiving of influence—as the yin-yang of the universe. There cannot be yang without yin or yin without yang.

Whitehead believes this twofold dynamic is at work throughout the universe, because human experience is an expression of, not an exception to, what is happening in other creatures. Many people find this plausible when they consider other living beings, especially if they have companion animals or if they are shaped by evolutionary biology. Dogs and cats, fish and birds, and even living cells seem to receive influences and also respond to them.

For many people, Whitehead's view becomes less plausible when it comes to inorganic matter. Mountains and billiard balls, they say, do not feel! Here critics of Whitehead typically find one of three alternatives more plausible. One is to say that the world is divided into two kinds of reality: things that feel and things that do not (dualism). A second alternative is to say that subjective states of living beings are ultimately reducible to things that do not feel (reductionism). A third is to say that subjective states are emergent properties that somehow arise out of complex combinations of unfeeling realities (emergentism). The net effect of these three responses is to say that when it comes to inorganic realities, there is only one kind of power and thus one kind of energy: *causal power.*

Whitehead's philosophy is a rejection of this idea. He believes there is something like receptive power all the way down. Here he is influenced by early quantum mechanics and its suggestion that the true building blocks of apparently lifeless objects are momentary energy events that arise and then perish within the flash of an eye. He believes the evidence suggests that these energy events within the depths of atoms both influence other energy events (causal power) and receive influences from other events (receptive power), and that the act of receiving involves a small portion of creativity, on the basis of which the precise details of the response cannot be predicted in advance. This is how he interprets the principle of indeterminacy. In short, Whitehead gives us new eyes for billiard balls! When one billiard ball collides with another and effects a change in the other, the first billiard ball exerts *causal power* on the other ball. And when the second billiard ball absorbs the impact and moves in a certain direction, it— or the energy events that compose it—exemplifies *receptive power*. Here again, causation and reception are complementary. Here again we have yin and yang.

Of course some people limit the word *power* to effective power. They say that the second billiard ball lacks any power at all and that only the first billiard ball has power. But the truth of the matter—indeed the truth *within* matter—is that there must be an exchange of energy in order for the second billiard ball to move. This means the second billiard must be able to receive energy from the first billiard ball. This capacity to receive energy from something else is an example of receptive power at the material level.

I have said that the kind of energy within atoms is only one kind of energy. For example, there is also the kind of energy that is exchanged between people emotionally and intellectually in acts of communication. Receptive power is especially important at this intersubjective level. Consider a simple act of communication. When one person shares a story with another, her aim is to exercise effective power. She wants to effect a change in the listener—that is, to be heard. It is only when her friend listens to her words, creatively synthesizing them into her own experience, that the communication is complete. The listener exemplifies receptive power. The act of communication depends not only on the effectiveness of the speaker but also on the willingness of the listener *to receive* what is heard. The feminist writer Nelle Morton was important for noting that this willingness to receive, so often neglected in more patriarchal understandings of power, has a palpable presence of its own. This means that receptive power has a kind of effective power as well. When other people listen to us it makes a difference in our lives. They hear us into speech.

Sometimes the power to receive is much more impressive than the power to effect change. When we meet a person who has a very wide mind

or a very open heart, that person has a kind of power that is often much
more impressive than mere brute force. Buddhists might call it wisdom
and compassion. In their wisdom these people can internalize many ideas
without reducing them to one idea. In their compassion they are able to
receive the sufferings of others and to share in the sufferings without being
overwhelmed by them. They have receptive hearts. To be sure, the power
they reveal to others is impressive and sometimes causes a change in others. Their power to receive is itself moving. But the power that moves us is
not that of coercion or impact. It is their willingness to be touched.

## God as Deep Listening

I have suggested that God has energy. I use the word *God* with some hesita-
tion, because it often suggests an agent standing above the universe who
acts upon the universe in a unilateral way. This agent is imagined on the
analogy of the subject of a sentence to whom predicates are then added,
such as acting in the world. God is the unmoved mover.

Many process thinkers reject this image of "God" because it presents
insuperable problems for theodicy. If God can move things in a unilateral
way, why doesn't he move the world toward compassion and justice? Why
would God choose to stand behind the world, limiting his own power?
Traditional theists respond with a free-will defense. God limits his power
in order to offer the world—or human life—the gift of freedom. Process
thinkers are uncomfortable with this defense. One source of discomfort
lies in the fact that there is so much violence in creation quite apart from
human freedom, of which predator-prey relations are an example. Roman-
ticists might say that as the rabbit chases the fox the rabbit "gives itself"
to the fox in order to help maintain a "circle of life." But process thinkers
doubt that the rabbit would agree. We do indeed live in a creative universe,
and predator-prey relations are one result of this creativity. But we also live
in a violent universe, and some of this violence has nothing to do with a
misuse of human freedom. This leads process thinkers to develop panentheistic approaches to God.

The word *panentheism* literally means everything-within-God. Many
Western panentheists assume that the real point of panentheism is that
God or God's Spirit is within the world. But the word itself suggests a
different point of view. It says that the world is immanent within God.
Whitehead speaks of this immanence as *the consequent nature of God*, which
can be described in two complementary ways. On the one hand, it is God's

ongoing activity of feeling the feelings, or prehending the prehensions, of the many events in the world as they unfold in time. It is divine empathy or deep listening. On the other hand, it is the many events of the world itself as they are continuously gathered into the unity of the divine life. It is the universe as a whole. The consequent nature of God resists any depiction in subject-predicate terms, because the subject is its predicates. It is the many of the universe becoming one in the divine life without losing their diversity.

Of course, in Whitehead the consequent nature of God is but one aspect of divinity. There are two additional aspects: technically called the *primordial nature* and the *superjective nature*. The primordial nature is God as a reservoir of infinite potentialities that may or may not be actualized in the universe, and it is a timeless activity of prehending those potentialities. This is God as the divine mind of the universe. The superjective nature is the presence of God as an indwelling lure within each creature by which the creature feels beckoned to realize its potential relative to the situation at hand. Technically speaking, this involves a weaving together of potentialities within the primordial nature and understanding gained from the consequent nature so that God can provide relevant potentials to each new moment of experience through "initial aims."

As some readers may know, the first generation of process philosophers spent many years trying to reconcile these three aspects of God. Hartshorne offers an especially helpful approach, although he eliminates aspects of the primordial nature that were important to Whitehead. For purposes of this essay, however, it is not relevant to enter into those debates. What is more important is to take these two aspects of God and consider them as two ways in which the divine reality might be experienced.

A certain kind of mystic might experience the divine reality as a reservoir of pure potentiality that lies behind or beneath the whole of creation (the primordial nature). If she is a pure mathematician, then the types of pure potentiality that most attract her will be the sheer beauty of mathematical relations, considered apart from any application to the world. If she happens to be an actress, the types might be the sheer beauty of various forms of emotion, which can be experienced in human life but can also be considered apart from their embodiments. In each instance the person will be momentarily absorbed in a world beyond yet relevant to this world.

On the other hand, someone who thinks of God in more personal terms might experience the divine reality when she hears prayers and shares in the sufferings of others (the consequent nature as God's prehending of the universe) and who then responds by offering possibilities for bringing wholeness out of tragedy (the superjective nature). And one who does not have such a personal sense of God might experience the divine reality as the sheer interconnectedness of things, somehow gathered into the unity of a diverse whole that has beauty in its own right (the consequent nature as the universe itself). In the house of religious experience there are various rooms, many of which make sense from a Whiteheadian point of view.

Thus the question emerges: what kinds of religious experience might be especially relevant in our time, given the need to develop sustainable communities that embody harmony with the earth and harmony among people? The answer may be all of them, but I close with a word about three forms that are especially relevant to students and educators.

The first is *the prophetic imagination*. I borrow the term from Walter Brueggeman. This lies in being honest about the way things are in the world and then imagining the world as it ought to be and should be, in contrast to the way it is. Of course it is followed by acting in light of that hope. In the context of process thought this would be one way of responding to the divine call toward shalom or, as I speak of it, inclusive sustainability.

The second form of religious experience is *the receptive heart*. This lies in listening to the many voices of the world, human and nonhuman, with a sense of appreciation for their intrinsic value and their interconnectedness, God or no God. In the context of process thought it would be one way of participating in the consequent nature of God, even if a person does not believe in God. In our time the voice of Buddhism seems to speak quite deeply to this kind of experience.

The third form is *the critical mind*. This lies in analyzing the way things are, seeing through the pretenses and illusions as best one can, even when those illusions happen to be one's own. This is important for people who are tempted to turn our own most precious beliefs into ideologies by which we shut others out. In the context of process thought this would be one way of acknowledging the reality of possibility and its difference from actuality. Sometimes we hold beliefs about ourselves and others that we wish were true but are not true given the evidence.

One of the aims of liberal education is to help people become "whole people." One service that departments of religion and philosophy can offer students is to provide them with opportunities for all three forms of consciousness: the prophetic imagination, the receptive heart, and the critical mind. Sometimes we lean too heavily toward the critical mind, but in the last analysis all three are quite important. Indeed, from a process perspective all three are religious in their way, even if not called "religious." One way, but not the only way, that higher education can assist in the development of sustainable communities is to undertake hard-nosed research that provides guidance in an age of declining energy reserves. But another is to help people become more religious in the three senses. Along the way, the word *religion* never needs to be used. The word *sustainability* will suffice.

# CHAPTER 7

# "One More Stitch":
# Relational Productivity and Creative Energy

*Donna Bowman*

ato hitome   kono ichidande   yoga akeru
あと一目 この一段で 夜が明ける

One more stitch . . . one more row . . . Ah! It's dawn.

TATA AND TATAO

The first warning sign for many Americans of the current recession and its associated crises was the 2008 spike in oil prices. Suddenly everyone was talking about four-dollar-a-gallon gasoline. The prospect of the increasing cost of fuel raising the price of nearly everything—from food, to travel, to labor—frightened many people, who saw shortages, inflation, and a return to the dark days of the 1970s on the horizon. Gasoline prices subsided, but the larger problems facing the American economy were only beginning to become visible: uncollectable debt, failing financial institutions, lack of credit availability, bankrupt businesses, unemployment. Consumer spending declined as people began to worry about their personal safety nets, trying to hoard their cash for the rainy days on the horizon.

In times of economic hardship, Americans have always turned to handcrafts. Making useful or decorative items by hand makes sense under these conditions as a money-saving measure, of course, but also as an attempt to insulate ourselves from the volatile marketplace with gestures toward self-sufficiency. The Depression saw the growth of the sewing pattern industry. During America's two world wars, magazines and newspapers were filled with tips on do-it-yourself projects, and people "knit their bit" for the defense effort. The fifties pulled both ways at the crafting movement; women were supposed to be freed from sufficiency concerns by labor-saving devices and increased leisure time, but the decade also saw a backlash against convenience products and a desire on the part of consumers to be involved in traditional domestic arts, even if only to the extent of adding an egg to a cake mix. With the rise of the counterculture in the sixties came an interest in natural, communal, self-sufficient living and personal expression, which gave rise to an explosion in crafting. Inflation and the energy crisis in the seventies built on that movement as millions of middle-class families supported a burgeoning hobbies and crafts industry, from macramé, to woodworking, to crochet.

It is not surprising that people's habits change in times of insecurity. What is interesting about the production of items by hand—intended for personal use, for sharing and gift giving, or as a cottage industry—is its relationship to energy both in terms of large-scale issues of fuel and infrastructure and in terms of small-scale issues of individual habits and networking. In this essay I argue that the unique conditions of the early twenty-first-century crafting movement suggest insights about energy that have theological and philosophical consequences. In particular, the opportunities afforded by informational infrastructure combined with the creative potential of individuals can demonstrate a middle way between the thoroughly industrialized future envisioned in the fifties and the postapocalyptic, postindustrial future imagined by many at this time of emergency. Instead, process thought offers resources for understanding the contribution of personal energy inputs into a technological nexus that results in a sustainable transformation of the material world.

When the World Wide Web was in its infancy in the 1990s, many forecasters felt that, for better or for worse, humanity was moving away from existing primarily in the material world and closer to existing primarily in a world of information. In such a world, the disembodied mind would be as well equipped—or even better equipped—to navigate, perceive, decide, and act. Some, like Ray Kurzweil, looked forward to a future in which the mind is plugged directly into the network, experiencing virtually far more than could be experienced materially.[1] Others warned of a *Matrix*-like outcome in which we can become satisfied with virtual inputs, leading to neglect of the body and lack of care for our physical surroundings.[2] Theologians and sustainability advocates alike have made efforts in the last few decades to call us back to an appreciation of the role of embodiment in human experience and to our physical connectedness to the ecosystem. These movements seem to be threatened by the growing complexity and availability of virtual interaction. On the other hand, process theology has long worked within a model of mutual and essential interrelatedness in terms of "data." The Whiteheadian term *prehension* attempts to broaden the range of inputs into a developing entity, covering not only those acquired through sensory perception but also those acquired through direct, intuitive, or non-sensory means. Seen in this light, the increased scope and diversity of inputs made possible by computer-mediated interaction becomes an advantage for conscious creatures making decisions and creating reality out of the data available to them. The Internet may truly be the next evolutionary stage for the human mind, allowing our senses to extend worldwide and enabling the information, ideas, and memes we create to spread to an audience that is qualitatively larger than that reached by previous mass-media forms, with their high barriers to entry.

Surprisingly, however, the brain-in-a-jar possible future that both advocates and detractors foresee has not turned out to be the direction in which online networking has taken us. Perhaps the most astonishing development in the growth of Web-based interaction is the tight connection to material productivity. If this bond were better understood, those in the theological realm who fear losing an understanding of and appreciation for embodiment and physical resources might embrace the creative forces found in virtual communities, harnessing them to adorn and transform our material environment.

Process thinkers have been working since the middle of the last century to formulate an empowering theological vision for an increasingly centralized, technocratic, Foucauldian world. The efforts of John Cobb Jr., Catherine Keller, Jay McDaniel, and many others are aimed to reclaim power from a hierarchical traditional worldview (wherein God is quantitatively all-powerful, and any power wielded by creatures is granted to them through divine self-limitation) so that individuals and communities can envision their activity truly changing the world's present and future. Alfred North Whitehead and Charles Hartshorne's metaphysical and theological framework for this redefinition speaks of decision making by actual entities, instances of reality with the power to create themselves out of the data of their past. Because of this language, and because of theology's insufficient attention, historically speaking, to the concrete and material, much of the explication of reorganized power in the process worldview has focused on changes taking place in the psyche. The future created in this moment, becoming real as this moment passes into the past and opens up new possibilities for the future, is often seen as existential selves being built, bricolage-style, by decisive action or thought. But Whitehead sees the material world as the product of these same forces. As each nugget of actuality "decides" (cuts off other possibilities for its becoming), all of the matter/energy that makes up ourselves and our environment is carried forward, whether transformed or repeated, into the next moment. It is time for process thinkers to turn their attention to the theological aspects of material production and to consider the tight connection between the power of agency motivating so much of our revisioning work and the energy gathered, expended, captured, and made useful in the manufacturing process. I submit that the unexpected flourishing of handcrafts mediated by social networking tools can serve as a prime example of how power once thought to be concentrated in fewer and fewer hands because of technological advances (Marx's analysis of industrial capitalism's effect on access to the means of production) is actually becoming more widely available than ever before and, more importantly, is demonstrating the inherent personal power of production to a rapidly growing participatory audience.

Let me offer just a few examples of how online activity leads to a kind of real-world production that values personal energy over the mass production that is the hallmark of the industrial age. I believe that a postindustrial understanding of both energy and theology must take into account the potential, already being realized in many quarters, of the Internet's empowering of the individual to become a maker of physical objects.

Threadless.com (www.threadless.com) is a virtual community masquerading as a clothing company. The company employs no designers or salesmen. They do not have an advertising agency, and there are no professional photo shoots of models. Instead, ordinary people submit graphic designs and slogans to the Threadless website. Other ordinary people vote on which ones they would like to see on a T-shirt. The company makes limited runs of the winning designs, with new designs added every month, and the person who submits the winning design gets a $2,000 prize, plus a bonus if the run sells out and the tee is reprinted. People buy the T-shirts directly from the website and upload pictures of themselves wearing them to be used as product images. In mid-2008 the registered user base (designers, voters, and customers) on Threadless.com was seven hundred thousand. Sales in 2008 were estimated at over $30 million. Most importantly for our purposes, interaction on a website by a dedicated, diverse, voluntary community results in the production of real-world artistic creations. As Max Chafkin wrote in a profile of Threadless for *Inc.* magazine last year, "This idea goes against a basic principle that has been taught in business schools since the invention of mass production: Employees make stuff, and customers buy it."[3] Instead, consumers supply the creative energy by sub-

mitting information, and the company acts simply as an enabler, organizing that information and distilling it into fabric and ink. Thinking about
Threadless only a few years into its existence, Tim O'Reilly, the author
and entrepreneur who coined the term *Web 2.0* wrote the following on his
blog: "How far off is a future in which the creative economy overflows the
thin boundary that separates 'information' from 'stuff'?"[4]

Around the same time as Threadless was breaking big, in 2005, O'Reilly
and others started a magazine called *Make*. The project was inspired by
the burgeoning hacker community, a group whose history stretches back
to electronics and radio hobbyists but had developed a bad reputation in
the Internet age as destructive online vandals, largely due to oversimplified
media reporting. *Make* contains features and instructional articles about the
creativity of "makers," people who produce modified cases for their computer towers, or clothing with embedded LEDs, or shoes recycled from
thrift-store castoffs. The focus is on a new breed of do-it-yourselfer—the
kind who sees the technology around him as a resource for scavenging, remixing, repurposing, and expressing individuality. The print magazine has
been accompanied from the very beginning by an active blog site (www
.blog.makezine.com) that updates several times a day with links to project journals on blogs, instructions posted at how-to sites like Instructables.
com, and images of notable objects gathered from online photo sites like
Flickr. Since its inception *Make* has spawned a sister publication, *Craft*
(www.craftzine.com), devoted to twenty-first-century implementations
of domestic arts like sewing, knitting, crochet, and so on; an online storefront selling electronic hobby supplies and tools; a public television show
and video podcast that premiered in January 2009; and three huge annual
"Maker Faire" expos in San Francisco; Austin; and Newcastle, England. At
the latest Austin event, more than one thousand exhibitors and eighty-seven
thousand attendees took part. The idea for *Make* magazine is not new at all;
it is simply an updated version of *Popular Mechanics* or other publications
devoted to hobbyists that flourished in the 1950s. What is new is the use of
the Internet to disseminate inspiration and instruction—the overwhelming
sense created by immersion in this community that such creativity is nothing out of the ordinary, and the provision of tools making it easy to join in.

Almost since the creation of the World Wide Web in the early 1990s,
handcrafters have been selling their wares on eBay, the venerable online
auction site that has evolved into an all-purpose Internet storefront for
hundreds of thousands of home businesses all over the world. However,
because the primary purpose of eBay is the selling of auctionable items
(antiques and collectibles, for example), the site is not geared toward the

personal production of the items sold, but only the transfer of items produced and acquired in any number of ways. Etsy (www.etsy.com) began in mid-2005 as a space in which makers could create individual shops and sell handmade items of all kinds, from clothing to art to food. More than 200,000 people now sell their creations on Etsy, catering to a customer base of over 1.8 million users. Total sales through Etsy in 2008 topped $87 million. It is possible to discern a movement from the participation of users/consumers/community members in the creation of a company's merchandise (Threadless), through the promotion of do-it-yourself energy and production by media entities (*Make*), to the facilitation of individual crafters selling their items to customers on a destination site that both affords them individualized shop space and aggregates their impact with other similarly enterprising fellows.

As a final example, and the one that provides a framework for the theory and theology of personal energy to emerge in the remainder of this essay, consider a social networking and database site called Ravelry (www.ravelry .com). The brainchild of a programmer whose wife is a knitter, Ravelry .com began in 2007 as a way to connect knitters and crocheters with information about their hobbies. Websites with information on yarn and patterns were nothing new, of course—thousands littered the Web, and some had community features like forums. But the vision for Ravelry was different. By providing knitters and crocheters with organizational tools, such as a projects notebook where each crafted item can be displayed and a stash feature where yarn and fiber can be catalogued, Ravelry aimed to harness its own users as the contributors, editors, and remixers of the information they came for. As a user builds his notebook by linking projects to patterns and yarn and uploading photographs, the data set connected to that pattern and that yarn becomes richer. Others looking for something to make with a particular yarn can browse the project pages of everyone who has used it and connect immediately to the patterns they followed. Still others trying to decide what yarn to use to make a particular pattern can browse the project pages of everyone who has made it, seeing how it looks in different fibers, on different models, with different modifications. Thousands of user-created and -moderated forums gather members around particular designers, books, magazines, even single patterns, as well as pop culture phenomena, other hobbies, religious interests, and even meta-discussions on goings-on in other Ravelry groups. One typical activity for almost all groups is the knit-along or crochet-along, in which any number of far-flung members craft the same pattern at the same time, posting on their progress and engaging in mutual support.

Ravelry membership exceeded a quarter of a million in January 2009 and half a million ten months later. The site came late to the so-called knitting explosion of the late 1990s and early 2000s. However, its embrace of Web 2.0 principles and technologies has placed this part of the crafting movement on a completely different foundation from that on which the first wave of knitting's new popularity was built. Ravelry now catalyzes the investment of personal energy into crafting by organizing and directing it in a nonauthoritarian, evolving, user-controlled framework. To understand how this marriage of social networking, user-built content, and individual production represents a revolution for the postindustrial understanding of energy, consider the history of knitting information in the last century.[5]

The predominant way knitting information was distributed before 2000 was through books and magazines—*traditional print media*. Magazines could be bought on newsstands, purchased at local yarn stores (LYSes), or received via subscription. Books were heavily dependent on knitting magazines, in which they are advertised and reviewed and with which they often shared a publisher. The only way to measure the size of the knitting community was through sales of these physical objects.

*Online knitting "magazines"* started to appear in the early 2000s. These sites solicited design submissions from the general public and published them for free (supported by advertising). These sources of free patterns, presented with editorial and quality control, quickly became overwhelmingly popular among a young, "wired" group of knitters who previously had not been recognized as a large segment of the knitting population. One of the largest and best known, Knitty (www.knitty.com), racks up an average of 50,000 visits and 160,000 page views every day, with 1.5 million unique visitors a month.

At the same time, *Internet knitting businesses* came into existence, selling yarn, supplies, designs, and traditional knitting media. Some of these businesses were Internet extensions of brick-and-mortar shops and LYSes, but a new breed of Internet-only, mail-order-only knitting businesses emerged in the early 2000s. All of these businesses formed a natural pool of advertisers for the online knitting magazines.

The availability of designs and supplies online freed knitters from their LYSes and other brick-and-mortar craft stores in their local areas. This made it possible for *more people in more places* to become knitters—all the materials could be mail-ordered or accessed online. The online businesses grew and offered larger selections. Two of the largest Internet-only businesses (Knit Picks and Elann) started their own yarn labels, offering highly affordable yarn in a range of popular natural fibers and weights. Local yarn stores with online shopping portals made it possible for any buyer to get high-quality brand-name yarns. Getting knitting supplies online increasingly involved no compromise in terms of availability, selection, quality, or price. In fact, the range of materials available from online businesses quickly exceeded that in even the best-stocked brick-and-mortar outlet.

*Online knitting communities* began as Usenet groups and Listservs, evolved into bulletin boards, and then into forums and boards conceived as adjuncts to the online magazines and stores. These communities not only allowed knitters to share information and built brand loyalty, but they also made the scale of the on- and offline knitting community visible in a staggering way. Ravelry, the first custom-built knitting online community, signed up one hundred thousand subscribers in its first eight months and continues to attract five hundred new members a day while still in its beta period. Its elegant design and integration of community tools with personal organization and social networking functions attracted the notice of Web design and usability writers, and its sheer scale earned mention in the mainstream media, where a spate of "Hey, what's with all the knitters all of a sudden?" stories was immediately visible.

One result of this Web-catalyzed growth is that new knitters and previously offline knitters are *more quickly and thoroughly integrated* into the knitting community than had ever been possible before. Help, suggestions, advice, and inspiration is available twenty-four hours a day. The eternal question "What do I knit next?" now has a universe of answers limited only by one's remaining lifespan, and any conceivable material is available to order online. It seems logical to conclude that more people are being motivated to try knitting because of this visibility and availability and that a smaller percentage of those novices end up being unable to sustain their interest because of lack of support, paucity of supply, or lack of information. In other words, the Web 2.0 platform for the community means the community grows both more rapidly and more sustainably.

Another, perhaps less intuitive result is that *traditional media are made more viable*, not less, by this online growth. Patterns in magazines and books are now "advertised" by being included in the online pattern library of Ravelry; those attracted to them seek out the magazine or book that includes them. Every new print-media release generates online discussion that exposes more community members to the ink-on-paper content. Designers and books have fan groups and knit-alongs devoted to them. And of course magazines and books are advertised directly by their publishers on these websites, sold through online knitting businesses, and reviewed by podcasters and bloggers. The penetration of traditional media into the much larger and much more diverse knitting community that has been revealed and nurtured by Web 2.0 applications is far more thorough than could have been achieved previously.

And, finally, the boomerang effect of all this online activity is the *creation of real objects*: knitted items from scarves, to sweaters, to DNA models, to laptop covers. Fifteen years ago most people would have considered handmade objects to be endangered artifacts in an increasingly digital world. Instead, *there are more of these handcrafted items in the world today* than there would have been without the Internet, and likely many more per capita than there were in any pre-Internet-boom period. Of course, real people have employment with real money in online-dependent businesses that make and ship real products, and real books have been written and bought and read because the Internet made some knitting bloggers stars, and so on in all the usual ways that the online world catalyzes real-world activity. But Ravelry is an example of Web 2.0 unleashing the power of the individual to create and therefore modify his particular environment. In larger terms, the proliferation of creative examples means that "designer" and "inventor" are no longer labels reserved for the few—everyone can be a "maker." This movement has empowered thousands upon thousands to create and invent, then to share their knowledge with others, which in turn sets off a new round of creation and invention. Informal social pressure creates a kind of healthy online competition to present and photograph one's work beautifully and to document the process of making thoroughly and helpfully. This builds and enriches the resources to which the community can refer for help or inspiration.

As a result, the world at large becomes newly aware that the "good old days" of craft, creativity, care, and the handmade are far from gone (as anti-digital forces used to take pleasure in lamenting). In fact, they are experiencing a renaissance and robust growth that has *never been seen before—a proliferation of expertise, skill, and creativity that would never have been possible before the Internet.*

These results may surprise both futurists, who assume we are leaving matter behind for a superior virtual existence, and neo-Luddites, who fear that information technology will destroy the virtues of traditional practices and processes. It may encourage process thinkers whose mission is to uncover the world-creating power and freedom inherent in all of us, in resistance to false hierarchies of oppression. It appears that when rightly organized and presented, information can connect intuitively in the lives of individuals to the materials and the processes of making that those individuals value. It can even create that value by removing the barriers to individual material production that frustrate and stymie many of us.

Process theology spotlights the role of relationships in the production of reality. Every individual creates herself out of the relations that pertain in her past. This self-creation has clear analogues to artistic production, although the stereotype remains that artists are solitary, reclusive types. We recognize nonetheless that they create out of their influences—and more so today, when art has become self-consciously an act of remixing and repurposing materials found in the environment rather than an attempt at pure originality. The difference between art and craft has typically been understood as the difference between the creation of messages, reflections, and statements about the world, and the creation of useful and beautiful things within the world. Yet craft is increasingly seen as a tool for art and vice versa. For example, artists install knitted and crocheted sculptures in museums and galleries, while knitwear designers, once anonymous women directed toward the utilitarian, are placed in the same creative pantheon as couture fashion designers.

The virtue of this increasingly central role for craft in the realm of creative making is that craft has always been understood to value and preserve traditions. Relationality, which has sometimes been obscured in art's quest for the novel, is highlighted in craft, focused as it is on the maintenance of techniques and forms passed down by generations of practitioners. Yet the blurring of the lines between art and craft means that art finds itself more interested in connecting with those traditions, while craft strives more than ever before toward novelty. Thanks to this mutual interpenetration, both fields have begun to exemplify the process notion of creativity, a force for the production of reality through a process emphasizing continuity but open to novelty. William Power, a process thinker at the University of Georgia, prefers the term *productivity* to the Whiteheadian *creativity*, since it more accurately describes the momentum of this universal force, which does not create each new moment ex nihilo, but produces it out of the past, which is available to the present emerging moment as raw materials. The change of language is felicitous for this essay. While the term *creativity* connotes artistry but lacks a sense of temporal direction, the term *productivity* emphasizes the "quid *ab* quo," "this from that" of the universal process. What becomes is not the only term of interest in the productive equation; what is used to make it is equally significant.

As we contemplate the use of energy in our Western way of life, focusing naturally on our dependence on fossil fuels and the need for alternative and sustainable energy resources, we typically and understandably fix our gaze at the level of the energy grid and infrastructure. We think about how electric power plants can keep supplying energy to our homes and businesses. We wonder what will fuel our transportation network and what supply chain considerations will be necessary at a regional or national or international level. In this essay I have attempted to turn our attention to the promise of energy at a different level. The productive power of individuals is burgeoning at a time when the productive power of industry is teetering on the edge of decline.

Our concern about energy use and abuse often leads us to advocate changes of habit on the part of consumers. We saw in the gas price spike of 2008 that consumer behavior can change very quickly in response to a strong stimulus. With a few months of gasoline prices over three dollars per gallon both the vehicular-purchasing priorities and the driving priorities of the population shifted dramatically. However, such changes are unlikely to become habits, because they are merely quantitative reactions to a quantitatively fluctuating stimulus. If consumers reduce their driving or seek out fuel-efficient cars when gas prices are high, that effect will fade over time if prices return to their previous levels. Driving will increase, and miles per gallon will cease to be as crucial a factor in new car purchases. If the change precipitated by conditions of scarcity is quantitative (a reduction in consumption), then when the scarcity is perceived to have ended, the change will end as well. Those advocating wholesale change, therefore, sometimes find themselves in the uncomfortable position of hoping for the worst, because only then will people be motivated to alter their behavior.

However, the change of habit that we truly seek is one that will not wax and wane with the intensity of the pressure from the environment or economy. In short, we seek qualitative change. And here it appears that the crafting movement has the potential to outlast the insecurity and uncertainty of the last decade. If people turn to producing their own food, clothing, toys, furniture, or electronics because their own productivity is cheaper and more reliable than that of the industrial supply chain, that is a qualitative change of habit rather than a quantitative one. It is not reducing consumption; it is taking upon oneself a role in production. It is not negative (I will have less of what the consumer economy supplies), but positive (I will have more of what my own productivity supplies). And if the habit is acquired and is found to be enjoyable, it will persist even when the environmental conditions that sparked it abate. The material condition that governs the staying power of a qualitative change of habit is the availability of the resources that support that habit—in this case, craft supplies such as yarn and tools. There is no reason that these supplies should fluctuate in availability or affordability in the same rhythm as the conditions that initially drove the individual to adopt the practice.

More importantly, a crucial condition has changed in the last decade that supplies a new baseline for individual productivity. The availability of information online to instruct and support crafters at all levels is not just a quantitative change. As demonstrated above, it is a kind of organization that allows for relationality on a different scale and with greatly enhanced potential. Let me enumerate just a few of the ways Web 2.0 changes the game for these productive arts.

1. *The economy of free.* Manufacturers have been giving away free samples to get business forever—because it works. But the Web allows for a huge economy of free, because marginal and scaling costs are virtually nonexistent. Give away digital content (like e-books) to build an audience for the author's next book. Give away patterns or instructions either so others can make your design ("paying" you in reputation and visibility), or to build an audience for non-free content through the creation of a loyal following. Everyone can participate in the economy of free, thanks to the Web. Of course, crafting has always been a leader in the economy of free. Its entry into the Web means that the crucial resource—expertise and instruction—is always available at the exact point of need.

2. *Social networking.* Crafting is an inherently relational enterprise; we learn from each other, and we make for each other. Now those connections are built in to a virtual community that extends one's network far beyond our physical reach. This increases the number of people we can learn from, the amount of expertise to which we are in proximity, the longevity of tradition (think Web-accessible digital archives), the relevance of information (think advanced

searches and context-dependent filtering), and the opportunities for connection for each member.

3. *Just-in-time production.* This phrase has been an industrial mantra for decades. But of course mass production always depends on the market's ability to sustain the large and expensive infrastructure needed to produce when demand strikes. Many Web 2.0 organizations that harness individual creativity are cottage industries. Because what they are doing is the opposite of mass industrial production, their overhead is small: server space, an e-commerce account, connections to local artisans or small-scale manufacturing if the organization makes the product—or no production needs at all if the aim is to distribute the production widely among members. Without costly infrastructure, which always lags behind the state of the art thanks to the time it takes to build it and the usable lifespan it needs to achieve to recoup that investment, these means of production can respond nimbly to changing demand. Raw materials obtained for one purpose can be reallocated to another without penalty.

4. *Handmade and individual.* Because of increased corporate dominance of the Web in the late 1990s, many concluded that the Internet was no longer a place where individuals could have an identity that they controlled and nourished—just a selection of brand names and company memberships. The individual production of hand-crafted items catalyzed by the Web is the best argument against this doomsday scenario. Hundreds of thousands of people all over the world producing their own useful and beautiful things provide a potent counterbalance to the homogenization and consumerism that once threatened to sweep from offline into online life.

5. *Spinning electrons into fabric.* And back again comes the wave, from the Web, where information is organized and made available by thousands working in their own self-interest, to the physical world, where useful and beautiful things are being made. A process that starts online, with relationships creatively ordered and nurtured, ends with homes furnished, bodies warmed and adorned, senses awakened, and spirits inspired by the energy of individuals.

The history of modern industrial civilization demonstrates that as the energy available from the grid decreases, the energy of personal production increases. We find ourselves in a period of decline in grid energy; some believe that this decline is permanent. Therefore we see the energy invested by individuals in their own productive material enterprises on the increase. However, even if new sources of energy are found that will reverse the trend and increase the availability of grid energy, there is reason to believe that the converse of that historical truism is not a necessary corollary. If the energy of mass production increases, the level of personal productivity may not reverse course and go into decline. I have argued that because this personal productivity is a qualitative change of habit, rather than a quantitative change in an existing behavior, and because contemporary online systems create relationships that support and catalyze this productivity, it has the potential to continue through any fluctuations in industrial and economic conditions.

In conclusion, I would like to suggest two fundamental theological lessons drawn from my observations at the intersection of handcrafting and the Internet. First, *poesis of information can lead to poesis of matter.* Process theology asserts that human beings share with God the capacity to create reality. As the Greek word *poesis* connotes, our making abilities shape both the material sphere and the information sphere. Artisans shape matter into beautiful and useful objects; poets shape words and ideas into beautiful and rich communications. In the information age the making of material things took on a secondary importance for many, and communicative capacities were of primary concern in Western societies less and less based on manufacturing and agriculture. Yet, surprisingly, as the Internet reaches succeeding developmental milestones, we find that the connection between individuals producing physical objects on the one hand and the gathering and making available of information on the other has been reestablished and strengthened. But this personal productivity is not empowered by any old arrangement of information. In order for virtually shared thoughts to flower in the form of individually crafted objects, the online community must be carefully shaped to form relationships, tap the knowledge of the members, and enable the remixing of data in novel and creative ways.

In stating the second lesson, I perhaps risk hubris. Yet strengthened by the process theological embrace of evolutionary change and emergent properties, I dare to suggest that *the advent of online relationships represents the potential for a new stage in the* imitatio Dei. For process thinkers, the role of God in the coming-to-be of each instance of reality serves to widen the individual's limited view. Through my relationship to God, I am freed from the narrow scope presented by my particular position in space-time. God, whose ubiquity means that God is proximate to all locations in space and time, grants me a larger perspective and presents to me possibilities for my own creative action that are not visible from where I stand, putting me in relationship to those whom I would not otherwise encounter, including entities in the past and future. As we extend our relationships online, we now in our finite way can participate in that divine role, just as we participate in the divine power of creation. We can place ourselves and other in relationships beyond our physical space, sharing data and shaping the creation of far-flung lives. This power of relatedness beyond the boundaries of our embodiment has been present ever since the development of language. Yet the Internet signals a new expanse of potential relatedness and a new level of richness to the information that can be shared. As I have argued in this essay, such relations can foster an environment where personal energy produces poetic material outcomes.

# CHAPTER 8

# Energy, Ecology, and Intensive Alliance: Bringing Earth Back to Heaven

## *Luke B. Higgins*

It should probably come as no surprise that a certain instrumental, techno-scientific rationality continues to dominate the discourse in which we currently engage conversations around energy consumption and its environmental effects. After all, it was modern science that defined the very concept of energy upon which much of today's industrial civilization was built. Science defines *energy* as *the ability to do work*, expressed as a quantifiable, exchangeable unit. The utility of the scientific concept of energy derives from its capacity to abstract from particular processes and exchanges that occur throughout the natural world, reducing them to exchangeable units, or "currency."[1] The conceptual tools of science are an undeniable gift insofar as they have vastly expanded and enriched our capacity for understanding and relating to the world around us. And yet it seems equally clear that many of the civilizational habits that have led to the devastating environmental crisis we face today have their roots in a certain approach to nature pioneered and institutionalized by modern science.

Identifying and constructively addressing these shortcomings in the modern scientific paradigm has proved a formidable challenge in part because of the way institutional and disciplinary boundaries have evolved in our society. As philosopher of science Bruno Latour has so discerningly pointed out, thinkers within the social sciences or the humanities—those disciplines from which the most trenchant critiques of techno-scientific rationality come—tend to be limited in their "jurisdiction" to the realm of politics, culture, and human values. Questions of nature, on the other hand, are handed over the instrumental, "fact-seeking" methods of the hard sciences. In Latour's view, these kinds of institutional divides have perpetuated a kind of paralysis in political ecology.[2]

As early as the 1920s, philosopher Alfred North Whitehead offered a compelling diagnosis of the ills of the modern industrial age—ills that he understood to emerge in tandem with modern science but should not be identified with scientific conceptuality as such. For Whitehead, there is nothing in and of itself wrong with the strategic reductions of scientific abstractions; it is these abstractions that allow us to draw novel connections between phenomena that would seem otherwise unrelated. Where modern science gets into trouble is with the error of mistaking these abstractions for some underlying reality—what he calls the "Fallacy of Misplaced Concreteness." Once these abstractions become reified such that they seem to point to some fundamental material substance at the base of the world, a particular relationship of the "thinking subject" to that "material object" ensues—one that Whitehead aptly describes with the term *bifurcation*. This term describes the uniquely modern "schizophrenic" tendency to split off thinking, value-driven subjects from natural, material objects governed by facts such that a completely different conceptual apparatus is invoked to negotiate each side of this split.

In this essay I argue that the fallacy of misplaced concreteness and the ensuing reality of bifurcation continues to offer one of the best diagnoses of how and why modern (and postmodern) Western civilization has found it so difficult to view itself in an interdependent partnership with the larger ecosystem. In the case of energy, these fallacies can lead to the illusion that abstract, exchangeable units of energy constitute the most real (or at least the most valuable) thing about the diverse natural processes and exchanges in which they are understood to occur. Together, the first two laws of thermodynamics and Einstein's famous theorem $E = mc^2$ seem to impart a sense that literally everything in the cosmos can be reduced to a substratal currency of power. Energy then becomes conceived as some underlying substance that is simply "there" in natural systems, waiting to be harnessed for human ends. In other words, the logic of bifurcation takes over—one in which humans, as active, thinking subjects, become the sole creative agents who can impart value to what is no more than objectified, mute, malleable, matter.

It is true that the startling, paradigm-shifting insights of relativity theory, quantum mechanics, and other branches of contemporary physics have posed certain challenges to the particular reductions of an earlier scientific materialism focused on mass (i.e., Newton's classical dynamics). Quantum mechanics, in particular, breaks down any simplistic binary between a neutrally observing subject and a mechanistic, law-determined object. However, I would assert that—to a great extent anyway—humanity at large has experienced these advances as simply further evidence of our nearly miraculous power to manipulate matter and energy for our own ends. In this way, perhaps the overall impact of twentieth-century scientific and technological innovations around energy has not been to challenge, but actually to enhance—to the point of nearly divinizing—a certain uniquely (though not exclusively) modern definition of what it means to be human—that of *homo faber*, man the maker.[3] If energy is the capacity to *do* or *make*, and everything in the universe can be translated if not directly transformed into pure energy, then there is theoretically nothing in the material cosmos that cannot be reduced to a kind of "standing reserve" for humanity's projects of making and remaking the world.[4]

For both Whitehead and the partnership of Ilya Prigogene and Isabelle Stengers—two philosophers of science (Prigogine is a Nobel Prize–winning chemist) who have been influenced by Whitehead's critique and vision—the modern West's propensity toward "misplaced concreteness" and a logic of bifurcation is not unrelated to its inheritance of a particular theological construct—one in which humanity's relationship with the natural, material world is modeled on an omnipotent deity's relation of dominance to the creaturely realm.[5] Part of what I suggest in this essay is that the secularism that emerged with modernity (modern science in particular) may be best understood not as advance *beyond* the theological so much as an advance *within* a certain reigning "theo-logic." This theo-logic, having both biblical and ancient philosophical/theological roots, is one in which the human capacity for reason (and perhaps language as well) is identified with a transcendent creator God's possession of a "master code" over the fundamental makeup of creation. Modernity, then, is less a rejection of this theo-logic than an enhancement of it: no longer content to simply imitate God's transcendent knowledge over creation, humanity must instead appropriate it for itself, to the extent that a deity external to the human mind is no longer necessary. Part of what I argue in this essay is that in order to address what are often understood to be the errors of a modern, *secular* logic, we must root out the *theo-logic* that implicitly undergirds this very conceptual framework. In other words, a problem that has theological roots, however buried that theology may seem to be, may need to be addressed (at least in part) with a constructive theological solution.

Whitehead's and Prigogine/Stengers's critical analyses of modernity and the problematic theo-logic with which it is intertwined can reveal some of the conceptual limitations that have stymied both secular and religious attempts to address the environmental crisis, particularly around issues of energy consumption. If our dominant paradigm for energy is a substance-based one—something simply there to be utilized—the only questions that can be asked are those of the efficient management and maximization of our resources. Certainly these are important issues to address, and there is no doubt that—as thinkers like Amory Lovins remind us—the transition from an emphasis on increasing energy output to an emphasis on increasing efficiency of energy use is certainly a move in the right direction.[6] And yet these types of innovations do not seem to constitute, in and of themselves, an adequate solution to the larger crisis—particularly if greater efficiency does no more than empower higher levels of consumption. Instead, greater attention needs to be paid to the larger ecological chain of events within which energy consumption takes place. We can no longer afford to focus merely on the process of using or managing something we imagine is simply "there." Rather, we need to begin with a more subtle analysis of the particular sequences of concrete, "historical," ecological exchanges through which various forms of energy are exchanged and then fit our own creative practices and habits into those exchanges such that they flow with, rather than against, these processes.

In other words, the entire process of energy exchange needs to be understood in terms of a dynamic network of collective, cocreative partnerships. The natural world is not simply a storehouse for energy resources; it consists of a series of active, agential, creative exchanges that give rise to that so-called energy. Likewise, humanity does more than simply consume some *given* resource; like natural systems, we are, ourselves, actively cocreating particular events and products that feed back into the same larger system where they have to find an ecological niche alongside everything else. Energy theorist Janine Benyus talks about how our energy technology needs to increasingly model itself on biotic systems in which all by-products and waste are fed back into the system to be used for something else.[7] I would take this even further in saying that our technologies not only need to *imitate* natural systems, but they also need *to be* natural systems—in the sense that human cultural and technological processes need to "lie flat" on the plane of our natural ecology. In Michel Serres's terms, we need to transition from being mere "parasites" on our biosphere to weaving our own "creations" into the larger, holistic living system of "creation" as a whole.[8]

If a problematic, "bifurcated" theo-logic hubristically posits human beings as the sole possessors of a master code that gives us the license to extract the earth's energies in order to make or remake the world in our own image, might we conceive an alternative theo-logic that configures the relationships between the divine, the human, and the earth differently? Divinity (associated to various degrees with human techno-rationality) would no longer be understood in the role of mastery over creation's underlying energy-substance, but instead would be conceived as that which inspires dynamic, creative alliance-building between and among creatures in an ecological system. Whitehead's concept of "intensity" provides a way of talking about the immanent, divinely cultivated value toward which all processes of creative-ecological becoming strive. In this way, it may provide an alternative way of conceptualizing the kind of involvement God has in ecological processes such that the cosmos is no longer reductively bifurcated into active mental subjects (of which God is the highest) and passive material objects.

Whitehead sees at the heart of reality not some static, law-determined substance (whether conceived as matter or energy) but a process of ecological becoming set in motion by an immanent divine lure. *Intensity* describes that simultaneously relational and self-creative propulsion at the heart of every occasion stemming from its aim at an *integrative* value—*integrative* in the sense of being both emerging from and contributing to the complex ecology in which it finds itself. Instead of placing the divine in an externalized position of transcendence—the theo-logic that has become the dominant model for the human mind's relationship to the natural world—our interface with the divine occurs in and through the "intensive alliances" on which the vital exchanges of a diverse ecosystem are based. This new theological conceptuality would function as a springboard for the most important task we have before us—namely, the hard work of constructing workable and flexible alliances between human culture and civilization and nonhuman creatures and biotic systems.

Clearly, Whitehead and Prigogene/Stengers are not the only thinkers to have attributed the environmental crisis to a certain modern techno-scientific logic that dualistically separates subject from object, mind from matter.[9] Neither are they the only ones to have associated these tendencies in modern thought with a certain theological inheritance. Continental philosophers such as Heiddeger and Derrida have exposed the ontotheological roots of the modern metaphysics that, to a large extent, constituted the philosophical backbone of modern science and technology.[10] Feminist ecological theorist Carolyn Merchant has given one of the most penetrating accounts of the way early modern scientists modeled their inquisition of nature on a certain patriarchal theological injunction to dominate nature (coded as female) in order to recover an Edenic state in which nature is perfectly submissive to its human (male) masters.[11] Neither does Whitehead's alternative metaphysic constitute the only basis for an eco-theology that would challenge Christianity's complicity with human exploitation of the natural world. Most first-wave eco-theologies have been based on a stewardship model in which the divine command in Genesis to have dominion over the earth and its creatures is reinterpreted as an injunction to be an ethical steward or manager of the earth.

I believe, however, that the approach I offer here—based on Whitehead's and Prigogine/Stengers's philosophy—offers a number of advantages. Most significantly, the broader explanatory framework of bifurcation allows us to simultaneously diagnose problems with secular science, philosophy, and religion/theology, such that none of these are simplistically pitted against one or both of the others. The original French title of the book by Prigogine and Stengers that I am most heavily drawing on in this essay is *La Nouvelle Alliance*, translated as "The New Alliance." It reflects these two philosophers' hope that we can move beyond an older alliance that pits the human mind, allied with eternal, divine "being" (heaven) over against earthly, material becoming (earth), toward a newer alliance where dynamic, human creative faculties are allied with the dynamic creative energies of the earth. Bringing their vision together with a theological reading of Whitehead's concept of intensity, I offer the phrase *intensive alliance* as a new organizing metaphor for theological cosmology. A theology of intensive alliance moves beyond the subjection of the earth to a sovereign heaven (corresponding in modernity to nature subjected to a sovereign mind) toward one in which heaven is to be sought precisely in and through the dynamic, intensive flows of earthly becoming. Intensive alliance also signifies an alliance of the abstract, connective power of scientific conceptuality with the highest ideals (ethical and aesthetic) of religion and the humanities—one of the primary objectives of Whitehead's philosophy. A theology of intensive alliance recognizes that in order to be truly transformative, eco-theology needs to move beyond the bifurcated theo-logic that (arguably) continues to inform many stewardship-based models. Instead, the creative, earthly processes of becoming need to be understood as the very interface for listening to and engaging the divine.

In the next section of this essay, I seek to trace a certain historical progression in the way energy has been conceived in relation to divine and human capacities. It is in ancient Greek philosophy that we find the roots of this bifurcated theo-logic in which the true essence of earthly, material entities is located in an external, transcendent sphere. Although the human capacity for reason and language is identified in certain respects with this transcendent master code, for the ancients the practical tasks of human life are still mostly negotiated on the earthly, creaturely—rather than the heavenly—sphere. This changes with the transition into modernity and the birth of science: in the modern epoch the transcendent abstractions formerly reserved for contemplation of the heavenly sphere are applied ever more agentially and instrumentally to the creaturely realm of becoming—a move that denies the reality of creation's "animate" becoming in an unprecedented way. Modernity is thus described by Prigogine/Stengers as a "bringing of heaven to earth," a process that ultimately renders a deity apart from the human mind superfluous. I argue that as humanity increasingly sees itself in the quasi-divine position of master of creation's secrets, we become less and less able to engage our relationship with the divine in and through the dynamic partnerships that link our lives to that of our natural ecology.

In the final, primarily constructive section of this essay, I give a more detailed account of Whitehead's concept of "intensity" and how it might empower a new theo-logic that no longer places humanity in the position of an external, God-like master manipulator of nature. Instead, it may encourage us to come back down to the plane of creation so that we might once again cultivate our relationship with the living God in and through the varied alliances that constitute a healthy ecology. Intensive alliance with the natural world is not a demand for humanity to forsake its unique capacities for making and unmaking—technological or otherwise—but it *is* a call to integrate these actions within a larger, ecological process of making and unmaking that we are embedded within. Divinity is thus decoupled from the idea of a single, abstract code or master plan for creation—scientific *or* ethical-religious—and understood, rather, as that which inspires and enlivens all entities in the cosmos in their complex, intensive relationships of mutual becoming. This conception of the divine can open the way to a truly transformative eco-theology in which God's call is heard in and through our immediate interactions with our living ecology.

## From Ancient Energeia *to Modern Energy: Bringing Heaven to Earth*

> It is this instinctive conviction, vividly poised before the imagination,
> which is the motive power behind [scientific] research: that there is a
> secret, a secret which can be unveiled. . . . When we compare this tone
> of thought in Europe with the attitude of other civilizations when left
> to themselves, there seems to be but one source for its origin. It must
> have come from the medieval insistence on the rationality of God, con-
> ceived as with the personal energy of Jehovah and with the rationality
> of a Greek philosopher.
>
> ALFRED NORTH WHITEHEAD, *Science and the Modern World*

I have suggested that, explicitly or implicitly, a particular theo-logic has
permeated the modern approach to energy—one in which humanity increasingly sees itself as the possessor of a quasi-divine master code capable
of harnessing and controlling nature's powers. In this section of the essay,
I briefly trace the development of this conception from its Greek beginnings to its culmination in modernity and, arguably, even postmodernity.
In the above epigraph, Whitehead suggests that scientific research rests
on a unique faith that a discoverable secret lies behind the assembly of
creation. He attributes this conviction to the combined legacy of ancient
Greek philosophical cosmology and the Judeo-Christian notion of a personal creator—God. I argue here that these theological roots continue
to profoundly shape our assumptions about where and how God can be
involved in the natural energy exchanges of the material cosmos.

The etymology of our word *energy* can be traced back to the Greek
word *energeia*, a term that had an important role in Aristotle's philosophy.
For Aristotle, *energeia* described a unifying, self-directed force modeled
most perfectly and prototypically by the divine unmoved mover. While
divine *energeia* could be characterized as perfectly simple, immutable, and
self-related, these qualities were only imperfectly approximated in entities
of the earthly, sublunar realm of becoming. On the earthly plane, *energeia*
summed up the capacity for one's final cause to teleologically rule over
the materially diverse aspects of our being—for instance, the soul's command and organization of the body. Both Whitehead and Prigogine/
Stengers see in this ancient Greek cosmology the roots of the modern
idea that a divinely transcendent rationality, external and immutable, governs over the various structures of the natural, material cosmos. In Plato's
and Aristotle's philosophies, for perhaps the first time, nature's energetic

forces are seen not to possess their own "animate" agency and intelligence (to which one related via one's alliances with various spirits, gods, or quasi-divine powers); rather, these are seen to derive from a transcendent, rational sphere.

Despite the centrality of this higher sphere for the workings of the human mind and soul, for the ancients, day-to-day human existence was still largely negotiated within the lower, material sphere of change and becoming. Eternal, mathematical laws only directly applied to the divine sphere of the heavens and thus were primarily objects of spiritual contemplation rather than tools for practical application to our earthbound life. In the world of Plato and Aristotle, a firm, hierarchical distinction was in place between theoretical thinking—the question of *why* things happen—and technological activity—the manipulation of *how* things happen. In certain cases, mathematical insights might be used to "trick" nature into deviating from its divinely ordained course, but this was considered an inferior deployment of philosophical knowledge. Humanity's practical relationship to nature's energies was thus still a matter primarily negotiated on the earthly plane of becoming, a realm whose unruly tendencies would always lie at least partly outside the ordered governance of heaven.

As we know, early Christian theologies of creation wove together the Greek rational cosmologies of their day with Hebrew notions of a world created through the personal (and specifically *linguistic*) agency of a creator God. To perhaps oversimplify this synthesis, the divine rational forms of the Greeks functioned as the kind of plan or guiding logic through which God actively willed the cosmos into being. Humanity was granted a special position in this creation as the one creature made specifically in God's own image. While I cannot fully pursue the following line of thought in this essay, I suggest that the full significance of the dominion passage in Genesis can be understood only in the context of a link between the *imago dei*—understood in terms of humanity's rational and linguistic capacities—and God's specifically linguistic creation of the world. It would be interesting to historically investigate to what extent human dominion over creation has been theologically justified by a certain "logocentric" identification of human rationality and language with God's very language of creation.[12]

Nevertheless, as staunchly anthropocentric as premodern Christian cosmologies may have been, it seems that none of them understood the lived human reality as taking place separate from the ecological matrix of creation. Though humanity might occupy the top rung of a "great chain of being," it hasn't quite leapt off that ladder entirely.

With the birth of modern science a new way of configuring the relationship between heaven and earth, divine and human capacities, begins to
emerge. In their work *Order Out of Chaos*, Prigogine and Stengers build on
Whitehead's insights into the uniquely theological framing of early modern science.[13] While the stereotypical narrative tends to pit early scientists against the religious establishment of their day, the two philosophers
convincingly argue that the early scientists' quest for a world of predictable, quantifiable processes in many ways dovetailed with increasingly robust theological accounts of an omnipotent God's supernatural powers.
As Prigogine and Stengers point out, the emerging scientific community
found the hierarchical, Aristotelian world "too complex and qualitatively
differentiated to be mathematicized" and too dogmatic in its assertions of
"final causes" to inspire empirical research into the details of nature's "efficient causes."[14] What is less often talked about, however, is that certain
theological factions were *also* dissatisfied with this Aristotelian cosmology,
finding it to leave the active, living processes of earthly beings too far outside the direct control of God's sovereignty.

Thus, an ironic convergence took place between physicists' newly
emerging desire to understand not just the why but the how of creation's
processes—as well as actively intervene into those processes—and a
newly emerging theological commitment to a type of divine power that
did not just provide a passive model for the earthly realm but also actively dominated every detail of its operations.[15] As I have made clear, the
moderns largely inherited from Greek philosophy their conviction that
the world was governed by a rationality lying outside of itself. However,
for the Greeks, nature's temporal becoming was fundamentally alien to
the eternal, divine realm and thus always at least partially outside of its
auspices. The early modern physicists, on the other hand, boldly asserted
that there was nothing on earth—no process, movement, or change—
that could not be precisely mapped through the lens of this divine intelligence. The world could be seen as a single, albeit complex, mechanism,
each portion of which was lawfully and deterministically positioned in relation to every other portion. Everything one would need to know about
the universe was—theoretically, at least—given all at once in each and every segment of reality. The eternal mathematical principles previously reserved for Aristotle's heaven were now directly brought to bear on earthly
processes—in Prigogine and Stengers's words, *heaven was brought to earth*.[16]
Theologically, this "discovery" of nature's orderly functioning was able to
serve as evidence of the world's utter submission to its divine master, or as
Prigogine and Stengers put it, "The 'mechanized' nature of modern sci-

ence, created and ruled according to a plan that totally dominates it, but of which it is unaware, glorifies its creator, and was thus admirably suited to the needs of both theologians and the physicists."[17]

These changes came with a major paradigm shift in the relationship between *techne*—active human making—and *physis*—the natural law of creation. The newly emerging experimental method presented a radical challenge to the traditional Greek partition between the active manipulation of nature and its theoretical comprehension. For modern scientists, knowledge was to be attained not by passively contemplating nature but by actively framing it in such a way that it was forced to yield evidence of the abstract principles that had been theorized. Experimental manipulation was no longer a dishonest tricking of nature from its true course as it would have been considered by the ancient Greeks; it was a way of uncovering the secret of what nature *really and truly was*. Exposed to the ascetic rigor of the experimental method, there was no piece of creation that could not be reduced to its bare, mathematicized essence.

As Whitehead brilliantly explains in *Science and the Modern World*, this new way of framing the cosmos left certain explanatory gaps—namely, what to do with all of those aspects of experience that did not directly exhibit these eternal, static laws. If the only thing that counts as information is that which conforms to the "idealizations that guided the experiment," how are we to interpret all of those secondary phenomena—including thought, freedom, perception, value—that cannot be directly traced back to physics' fundamental principles?[18] A somewhat radical solution was provided by the moderns: to claim that these secondary qualities did not really exist. Or if they did exist, they did so in an entirely separate ontological sphere (the dualistic metaphysics of Descartes, for example)—one that human beings alone of all creatures had access to. Whitehead uses the term *bifurcation* to describe this uniquely modern dichotomization between matter and mind, nature and culture, objective primary phenomena and subjective secondary phenomena.

Although it remains outside the parameters of this essay to put forth a historically rigorous analysis of these correlations, I would submit, along with Prigogine and Stengers, that the early modern scientists' increasingly interventionist understanding of their research methods (literally prying loose creation's secrets) was deeply informed by a certain Christian theological inheritance—in particular, a God who brought together, in Whitehead's words, "the rationality of a Greek philosopher with the personal energy of Jehovah" (see the epigraph to this section). For the moderns, however, this was no longer a God to merely worship from afar, but one whose mastery of creation we were to emulate and whose secrets we were to appropriate for our own ends. In this model, an "alliance between *man*, situated midway between the divine order and the natural order, and *God*, the rational and intelligible legislator, the sovereign architect we have conceived in our own image," utterly trumped any other alliances humanity might have kept with the earthly, energetic forces of nature.[19] Knowledge of the secret laws, the master code by which God had actively assembled creation, served as a kind of commission to actively make and unmake creation in *our own* image.

A closer look at Newton's theory of gravity and the reception it had in its day might help us grasp the workings of this theo-logic in early modernity. As discussed above, the earlier Aristotelian concept of *energeia* explained the rational *telos* behind nature's processes; however, it applied primarily to the eternality of the heavens and only derivatively to the earthly realm. In contrast to this, Newton's theory of gravity attempted to prove that the motions of *both* heavenly and earthly bodies were determined by the *same* force. Newton's ambition was to reduce literally all physicochemical processes—including heat—to the mathematically determined relationship between mass and force. Knowledge of these relationships was understood as no less than a kind of direct access to God's own viewpoint on the world. Newton considered his theory of attraction as a definitive proof of the existence of God—and not a passive, deist God stereotypically ascribed to most early scientists, but an active God, whose agency was instrumentally responsible for coherently bonding together the very fabric of creation. Indeed, in its time Newton's theory of attraction was often explicitly interpreted as a kind of quasi-divine revelation, one that empowered humanity to take its rightful seat next to God. Prigogine and Stengers quote a contemporary of Newton: "Then the Word was made man, the Word of the seeing God Whom Plato revered, and He was called Newton. He came, and he revealed the principle supreme, Eternal, universal, One and unique as God himself. The worlds were hushed, he spoke. ATTRACTION. This word was the very word of creation."[20] In a manner of speaking, the words by which God had assembled the universe were now the very words yielded by man to dominate and control nature.

Despite the series of devastating blows dealt to Newton's ambitious vision for science by later breakthroughs in physics, Prigogine and Stengers argue that it is this "revelatory" understanding of scientific knowledge that has most deeply shaped science's fundamental understanding of its task. Physics' central categories of explanation—whether identified as universal attraction, field theory, elementary particles, or simply energy—continue

largely to be understood as the very "word of creation."[21] This theo-logic
posits science less as an open-ended dialogue with nature's own "animate"
agencies (the alternative vision advanced by Prigogine and Stengers) than
as a mastery of its "True" essence from "on high." As we know, science
soon came to distance itself from these early theological interpretations
of its work. I would submit, however, that science's diminishing use for an
explicitly religious deity may actually correlate with its increasingly thorough appropriation of a quasi-divine position for itself.

I submit that the theo-logic outlined above can account for some of
the most significant limitations in both secular and religious approaches
to our natural ecology. Insofar as the logic of bifurcation comes to define the relation between human culture and its natural, ecological context,
the "energies" experienced and exchanged in thought, culture, and even
religion become increasingly distanced from our conceptions of energy
as a natural material resource. The former are understood to be actively
cultivated through the interactions of historically particular, animate agencies, but the latter are understood to be merely there—a standing reserve
to be instrumentally utilized by human civilization. The energies of the
natural world thus come to have no inherent meaning, value, or history
outside of that which humans give to the idea. Natural history teaches us,
of course, that our various natural resources are the product of an incred-
ibly long and particular history of complex ecological interactions; fossil
fuels are perhaps the best example of this. And yet we find it difficult to
think of these processes as the interaction of agential subjects pursuing
a value immanent to their own reality; this logic is reserved for human
civilization alone.

As far as the hard sciences go, as I mentioned above, Newton's reductive materialism based on mass and force has been radically challenged
by advances in contemporary physics, including theories of indeterminate
quantum energy fields; a relative space-time continuum; and systems that
behave in strangely chaotic, "far from equilibrium" patterns. Prigogine
and Stengers draw particularly upon the latter for the central argument
of their book—that we need to move toward a less controlling, more dialogical conception of science locating itself *within* a temporal process of
becoming, with all its risks, rather than above it. It is not within the scope
of this essay to unpack the implications of these new sciences for modernity's reductive materialism. While certainly we can hope that these new
scientific paradigms will help transform our culture's habits and attitudes
toward its natural ecology, we also have to admit that they have so far failed
to do so. On the contrary, new developments in science and technology

of energy seem—if anything—to have enhanced humanity's sense of its quasi-divine position as maker and unmaker of creation. After all, these new technologies allow us not just to manipulate the relationship between force and mass as in Newton's time but also, as in the case of nuclear fission, to literally take apart matter at the seams—to dematerialize and re-materialize it for our own purposes.

This bifurcated theo-logic explains not just the conceptual limitations of science and technology, but also those of contemporary religious and theological approaches to the environment. Since modernity, religious institutions have drifted further and further toward one side of bifurcation's great divide—the side of thought, value, feeling, and culture. While nature may provide poetic fodder for theological metaphors, on a practical, day-to-day level, our interactions with our environment are ruled by instrumental, scientific fact. In the last few decades, an emergent movement of eco-theology has attempted to bring theological values to bear on the environmental crisis. Many of these eco-theologies (especially those based on the stewardship model), however, have largely continued to root themselves in the idea of a privileged alliance between humanity and its creator God. The major difference is that this alliance is interpreted to charge us with exceptional *responsibilities* toward creation rather than exceptional *privileges*.[22]

Stewardship eco-theologies still conceive of God as the architect of creation's assembly, our relationship to that God (as creatures made in God's image) continuing to offer us privileged access to the divine "logos" guiding creation's processes. Now, this "logos" is understood less as an abstract code to be used for manipulating creation than a kind of abstract wisdom intended to ethically guide our interactions with creation. However, the basic configuration of divinity, humanity, and nature's energetic processes is still the same; God's influence continues to take the form of a revelation from on high that provides the correct "key" to the management of nature. The major shortcoming of this theo-logic is that God is still identified with an abstract "master plan" to be applied *to* nature rather than a living inspiration to be sought in and through our dynamic partnerships *with* nature. The result is that despite its good intentions, Christianity continues to perpetuate the modern (and partly premodern) idea of a de-animated, disenchanted nature to be commanded from above. A theology such as this is incapable of calling us to listen for God's call in and through our immediate interactions with our living ecology, something a radically transformative eco-theology must do.

*From Energy Management to Intensive Alliance:*
*Bringing Earth Back to Heaven*

> The radical change in the outlook of modern science, the transition
> toward the temporal, the multiple, may be viewed as a reversal of
> the movement that brought Aristotle's heaven to earth. Now we are
> bringing earth to heaven. We are discovering the primacy of time and
> change, from the level of elementary particles to cosmological models.
> Both at the macroscopic and microscopic levels, the natural sciences
> are thus ridding themselves of a conception of objective reality that
> implied that novelty and diversity had to be denied in the name of im-
> mutable laws.
>
> ILYA PRIGOGINE AND ISABELLE STENGERS, *Order Out of Chaos*

I have argued that a certain, sometimes overlooked, collusion between
the emerging abstract conceptuality of modern science and a particular
Christian theology resulted in the dominance of an especially problematic
model for conceptualizing humanity's relationship to its natural environment. This model has privileged humanity's alliance with an eternal, transcendent, divine sphere (whether or not it is explicitly recognized as such)
at the expense of its multiple alliances with other living processes of this
earth. Prigogine and Stengers's phrase "bringing heaven to earth" captures
the sense in which science has understood its conceptual tools (implicitly
or explicitly) as a kind of quasi-divine master code capable of capturing
creation's fundamental essence. In the case of energy, this "lens of heaven"
reduces creation's natural, fluctuating processes to an underlying currency
of energy with no inherent value until it is "plugged into" human civilization. The logic of "bifurcation" that emerges here prevents us from attrib-
uting to nature's energetic processes any of the animation and agency that
we would apply to our own human, cultural, and social exchanges. Nature
and culture take place on different levels of reality and are guided by fundamentally different principles.[23] It thus becomes all too easy to simply
bracket out the deeply complex, historical, ecological processes through
which nature's energies came to be in the first place, as well as the impact
that our *own* creations make on these processes.

Along with Prigogine and Stengers, I believe the time has come to
move in a direction converse to that of modernity and *bring earth back to*
*heaven*. In other words, the dynamic processes of earth's becoming should
be the very interface through which God is sought. Instead of appointing

us quasi-divine managers of creation's underlying energetic essence, God's role is to inspire in us a deeper responsiveness to our living ecology. God is to be encountered in *all* of nature's processes as that immanent lure that enables creation's integrative, ecological becoming. In this section of the essay, I show how Whitehead's philosophy, particularly his concept of *intensity*, can help us resist the logic of bifurcation and reenvision the terms through which we have traditionally related divinity and energy. Intensity gives us a way of talking about a divinely inspired, dynamic relationality at the heart of the cosmos that can never be reduced to some uniform substance to be controlled from above.

Unlike many other nineteenth- and twentieth-century philosophers, Whitehead—with Prigogine and Stengers following in his footsteps—does not simply oppose philosophy's conceptuality to that of science. Rather, philosophy is charged with providing a speculative framework that is capable of coordinating insights from a wide range of disciplines, science included. Philosophy's supreme task, for Whitehead, is to bridge science's systematic, detail-oriented attention to "brute matters of fact" with religion's pursuit of our highest ideals. As I explained in the first section of this essay, the problem with science lies not in its power of abstract reduction but with the misperception that these abstractions can capture some static, fundamental essence at the heart of the cosmos. When we mistake abstractions for something real, we fall into what Whitehead calls "the fallacy of misplaced concreteness."

Whitehead's reading of Newton's *Scholium* in *Process and Reality* provides a relevant case in point. While he appreciates the *Scholium*'s "immensely able statement of details" that can be "thoroughly trusted for the deduction of truths at the same level of abstraction as itself," Whitehead sees it lacking an awareness of the "limits of its own application," in this way falling victim to the fallacy of misplaced concreteness.[24] Like Prigogine and Stengers after him, Whitehead characterizes these limitations in Newton's thought in specifically theological terms. Newton's static cosmos—in which every piece is perfectly and mechanically "ready-made" for every other piece—rests on the supposition of a God whose eternal, rational nature actively penetrates every aspect of the world. "For the *Scholium*, nature is merely, and completely, *there*, externally designed and obedient."[25] For Whitehead, humanity's appropriation of a quasi-divine lens on creation is a key component of this logic of bifurcation that refuses to recognize in nature any of the animate, self-generative agency that humanity attributes to its own intellectual and cultural processes.

The project of "bringing earth back to heaven" requires us to reconceptualize the divine in such a way that it is not alienated from but woven into the dynamic processes of becoming that characterize both nature and culture, a task to which Whitehead's philosophy is uniquely suited. Whitehead identifies at the heart of reality not simply located, substantial entities governed by *external laws* but rather self-creative events ("actual occasions") that pursue *values* inspired by a divinity that is—at least partly—immanent to themselves. In this schema, mentality is not associated with some quasidivine ability to know and/or supervise the physical materiality of creation from above; rather, it is woven, on some basic level, into the very makeup of *every* occasion in the cosmos. Each and every actual occasion in the universe is constituted by both a physical and a mental pole; the former constitutes the past "datum" that the occasion must integrate within itself (the *what* of the occasion), and the latter constitutes the open possibilities for the negotiation of this integration (the *how* of the occasion). Even the simplest bit of inorganic matter or swath of empty space *experiences* its world and is—to some extent—an agent of its own becoming. Although the mental pole will play a more extensive role in "higher grade" living occasions (the highest of which we are aware is consciousness), it is present in at least some form in every form of actuality.[26]

For Whitehead, the notion of "intensity" captures the integrative value being sought in the becoming of each occasion. The creative spark that sets that self-creative process of becoming in motion is a divine lure (often called the "initial aim") that opens relevant possibilities for that occasion. This lure aims at intensity of satisfaction, which is achieved through a particular kind of integration of past influences, one that allows them to work in creative *contrast* with one another rather than mutually inhibiting one another. In a certain manner of speaking, then, intensity feeds off of contrast.

As I have explained, Newton's cosmos was one where no becoming, no evolution could take place; its structure was given once and for all by God's reigning intelligence. Whitehead's cosmos, on the other hand, is one whose complex structures of order are *emergent from* a long, complex history of this very process of integrative becoming I have described. Actual occasions employ certain patterns ("eternal objects") in their prehension of one another, patterns that get repeated and extended throughout the universe. Although these patterns give the world a significant degree of consistency and continuity, there is nothing absolute about them; their repetition always happens with a slight (or sometimes dramatic) difference. Most importantly, these patterns are not transcendently dictated from above; they are divinely inspired (arising out of the "primordial nature of God"), but their emergence is grounded in an open-ended, relational history that could always have unfolded differently. Thus, even something as seemingly fundamental to our universe as electromagnetic laws are seen by White-head as a creative *achievement* of our particular cosmic epoch rather than the lawful execution of some eternally transcendent set of forms.

The evolution of the cosmos toward complexly structured orders can-not be understood apart from the seeking after intensity that happens at the heart of each occasion. Intensity is a product of contrast, and contrasts are made increasingly possible through the evolution of a complexly ordered cosmos. "The heightening of intensity *arises from order* such that the mul-tiplicity of components in the nexus can enter explicit feelings as contrasts and are not dismissed into negative prehensions as incompatibilities."[27] Depth of intensity occurs when an occasion's contrasts are able to strike a balance between a certain "narrowness"—an ordered consistency in its immediate predecessors—and a certain "width"—a diverse background.[28] These conditions are best achieved in an environment that combines di-versity, a certain amount of order (social relations), and a certain amount of chaos (nonsocial relations). In a manner of speaking, then, it is the seeking after intensity that is behind the emergence of complex ecologies in our cosmos.

This aim at intensity, for Whitehead, is first and foremost grounded in God's very nature. God's primordial nature inspires occasions to generate an intensity in their relational becoming that is then received into God's consequent nature. God's

> aim for it [the occasion] is depth of satisfaction in an intermediate step towards the fulfillment of his own being. His tenderness is directed towards each actual occasion as it arises.
>
> Thus God's purpose in the creative advance is the evocation of in-tensities. The evocation of societies is purely subsidiary to this abso-lute end.[29]

In short, the complex ecology of the cosmos comes about as a direct re-sult of God's pursuit of intensity. It is important to understand that this divine "appetition" that initiates the "concrescence" of each occasion has no deterministic power over the world's becoming. No single set of di-vine laws—ethical or scientific—commands over creation. Divine value is only emergent *within* the particular configuration of relationships that each occasion finds itself embedded within; thus, it is only *within* the flux of creation's matrix that God's call can be discerned.[30] In terms of earth's living creatures, God's cultivation of intensity in them aims at deepening the dynamic, interdependent relationships that constitute diverse, living ecologies.

The notion of intensity relates divinity to the dynamic processes of creation in a way that the concept of "energy" never could. Whitehead understands the modern scientific concept of energy as a useful tool for abstractly quantifying the patterned, vectorlike flow of prehensions among occasions within a particular nexus of occasions. In his words, "The physical theory of the structural flow of energy has to do with the transmission of simple physical feelings from individual actuality to individual actuality."[31] This abstract conceptuality, however, does not capture some *given essence* at the heart of material reality; to construe it this way would be to fall into the fallacy of misplaced concreteness. If we want to conceptualize the underlying, generative "appetition" underlying energy's vectorlike flow, what we need instead is the concept of *intensity*.

Unlike energy, intensity does not suggest some underlying, exchangeable force to be harnessed and/or manipulated from above. Intensity is not merely "there"; it is agentially produced from point to point, moment to moment, event to event, in and through creation's pulsations of becoming. Intensity reminds us that the power to do work is never merely a given resource in creation; it is always and only cultivated through complex, ecological, interactions, having their own particular histories, all of which are inspired, to some extent, by the divine lure. While the term energy can be easily conceived as no more than a resource for consumption, intensity frames our alliances with nonhuman entities and ecosystems as a process of cocreative becoming in which we must responsibly participate.

Instead of understanding our civilization's creations as fueled by "raw" energy, we can begin to think of them as products of various intensive alliances with creation. In this way, we can no longer avoid situating them within the larger, ecological processes of intensive becoming that are happening everywhere, all the time. In short, Whitehead's understanding of intensity helps to bridge modernity's great bifurcation: creation's energies and human energies are never two separate kinds of things but instead interact on a single ecological plane. Energy resources must be honored as living, animate agencies with their own histories and their own inherent value. Energy is not something to be merely managed, but something to be co-cultivated through intensive alliances with other nonhuman agencies in a larger ecology. In short, a more animate, enchanted understanding of energy may begin to emerge, one that does not simply oppose physicality to mentality or situate the divine in a purely external, transcendent position.

Whitehead's framework can help move eco-theology beyond its reliance on the idea of humanity's privileged alliance with the God who holds the master code of creation. The divine cannot and does not secure us in a transcendent position outside of the flux of creation's processes. To encounter God we must get our hands dirty; we must engage in those always risky alliances that bind our own flourishing to the flourishing of other beings. Divinity does not simply offer us ethical insight into our managerial tasks; it meets us in and through our intensive alliances with creation's living forces. In short, there is no alliance with the divine that can come at the expense of alliances with the other nonhuman agencies with which we share this ecosystem. In this Whiteheadian-inspired approach, God can be found only in and through the always shifting, always flowing ecological interactions that constitute life on this planet. Heaven can be sought only through the lens of the earth's becoming.

# CHAPTER 9

# "Go Big or Go Home": A Critique of the Western Concept of Energy/ Power and a Theological Alternative

## Oz Lorentzen

He plants a fir, and the rain makes it grow. Then it becomes
something for a man to burn, so he takes one of them and warms
himself; he also makes a fire to bake bread. He also makes a god
and worships it. . . . Half of it he burns in the fire; over this half he
eats meat as he roasts a roast and is satisfied. He also warms himself
and says, "Aha! I am warm, I have seen the fire." But the rest of it
he makes into a god. . . . he also prays to it and says, "Deliver me,
for you are my god." They do not know, nor do they understand,
for He has smeared over their eyes so that they cannot see and their
hearts so that they cannot comprehend. *No one recalls, nor is there
knowledge or understanding to say, "I have burned half of it in the fire and
also have baked bread over its coals I roast meat and eat it Then I make
the rest of it into an abomination, I fall down before a block of wood!"* He
feeds on ashes; a deceived heart has turned him aside. And he cannot
deliver himself, nor say, "Is there not a lie in my right hand?"

ISAIAH

The universe is a communion of subjects, not a collection of objects.

THOMAS BERRY

Discussions of energy typically include two different forms of energy. One
is directed toward being or becoming, often called psychic or spiritual.
This has to do with *potentia*, looking at the ability to actualize the intrinsic
or essential properties of an entity. The other is directed toward action or
doing and is often called physical. This has to do with the ability to work,
to accomplish tasks in the external world. While both forms of energy are
part of any concept of energy, typically one or the other is the focal point
and subsumes the other, either intentionally or by default. I suggest that

an adequate theory of energy should explicitly account for both. The history of the concept of energy in the West can be seen as a move from the role of *potentia* to that of work.[1] This corresponds to the ascendancy of the materialist scientific worldview in the West.

The basic error in Western approaches to energy comes from the materialist conviction that underwrites Western rational and cultural consciousness.[2] From this position, energy configured as a material property and directed toward material ends, the various approaches to energy recreate the same oversight/error. Energy is seen as tool, a means that is able to be utilized toward arbitrary ends. Energy is dislocated from the human (from a value system), allowing for mastery over the nonhuman, and it is directed toward doing things. Work, accomplishing things, is the essential feature of energy. This vision leads to a culture that is dependent on natural resources (human, animal, and nonliving) and develops a false consciousness, since it seeks to ignore the consequences of the second law of thermal dynamics (entropy) by, as Thomas Berry puts it, seeking to avoid the payment for this type of use of natural resources.[3] This dependency—cloaked as "mastery" and exacerbated by cultural denial—makes the human a slave to the systems built on these understandings of energy.[4]

Instead of materialism, I propose a fundamentally different beginning point, one that has implications for our approach to energy and our use of energy. My beginning point is structured around personalism—that is, I propose that the best way to approach the external world is by realizing that its source, its underlying reality, its cause, is personal. This conviction of the personal nature of substantial reality depends upon two theological considerations:[5] the first is theism, a belief in God as person; the second, is incarnation, an understanding of the divine that sees a more complete revelation of the divine in the historical person of Jesus of Nazareth. Since I take it as a truism that what is essential to personhood is who a person is, not how productive he or she is, the main difference pertinent to energy between the two approaches can be stated as *a focus on being, not doing.*[6] In this model, being is primary and doing comes out of (the right) being.

My theological presuppositions view the term *God* as limit language for the boundaries and necessities of human thought. Classically we could point to Aristotle's "unmoved mover"—that is, to the persistent notion that the idea of an infinite chain of events leading to the present moment as an account for our present lived/felt reality lacks both logical and existential satisfaction. When this ultimate term, a necessary condition for (Western) thought, is paired with a faith claim about the personal, involved, engaged nature of God—that is, the God of Abraham, Isaac, and Jacob—this basic condition for accounting for present existence has a decidedly personal flavor. Thus far, the weight of the three monotheisms (Judaism, Christianity, and Islam) can be appropriated for my position of personalism. Secondly, from an explicitly Christian doctrinal stance, the claim that in the historical figure of Jesus of Nazareth we see two complete natures (unmingled yet conjoined)—the divine and the human—in one person makes the Christian case for personalism even stronger. That is, if the divine is such that it can be "married" to the human through a shared personhood, then person must be substantial (or essential) to the divine nature.

These two theological considerations lead to several guidelines for an approach to energy and the energy crisis. First, as stated above, the focus moves away from the external to the internal. The question is not what but who, not quantity but quality, not charisma but character, not show but substance, and energy (important energy) is seen as that which aids in this development. Second, and related, energy is necessarily directed toward the freedom and opportunity to develop into a full humanness; energy is thus always antithetical to human dependency. Third, we see that energy is characterized less by motion and action and more by rest, where the greatest power is characterized by a (perceived) weakness. And finally, fourth, this includes an irreducible care and concern for the other, especially those who are marginalized or placed at risk by existing cultural approaches to energy/power, and here we see that energy is characterized by working toward the common good. In this essay I develop these theological considerations more fully before returning to these general guidelines as a sketch for an alternative approach to energy.

While many different strands of this theological approach could be developed, the one I believe is most germane to this conference and topic comes through a reading of the Sabbath. What does the Sabbath contribute to our understanding of divine energy and the human convergence with that energy as modeled by the incarnation and made accessible through the death and resurrection of the divine human?

This reading of the Sabbath will function through the theological strategy of metonymical pressuring of three key biblical texts:[7] Genesis 2:2, "God rested from all his works"; John 5:17, "My Father is working until now and I myself am working" (Jesus referring to his healing on the Sabbath); and Hebrews 4:9–10, "So there remains a Sabbath rest for the people of God. For the one who has entered His rest has himself also rested from his works, as God did from His."[8]

God continues to work, and the work of the Son (the divine human) is synonymous with that work. Yet, since the beginning of time, God rests, and His central command to His people is to enter that rest: to keep the Sabbath day holy![9] This rest is a cessation of work (of one's own work), but not a nonworking (the father works until now . . .). The son, in a like manner, offers rest to the weary and heavy-laden: *through* a yoke! (Matthew 11:28–29). The weariness of labor is that which rest answers to; weariness is not, however, an essential component of "work," since the divine working is characterized as a rest, *sabbath* (literally, a cessation). One could suggest that the entropy, the residue of spent energy, that is central to the materialist approach to energy (and thus the legal requirement of Sabbath rest for humans and the land) is not a factor in the divine (or personalist approach to) energy, now open to humans to share. What we need rest from and renewal for in human work is no longer part of work inside the working rest of the divine.

I think the way to understand, and do justice to, these two components is to appeal to the "epistemology of the cross": as personified in Jesus and explicated by Paul (primarily in the Corinthians correspondence).[10] This epistemology approaches the real through a paradox or inversion: the weak are strong, the foolish are wise, the things that are not set aside, the things that are . . . (I Corinthians 1:18–31) That is, the cross portrays the divine as a surd within the system of human cultural thought. It is a backward world, where losing is keeping, the first are last, the rulers serve, and where power is manifest as weakness.

How are we to make sense of this? First off, we can't! That is part of the point that "the flesh profits nothing" (John 6:63). In order to make sense of this, we must suffer the surrender and death of everything that currently invests our world/universe with order and meaning. This is, paradoxically, not a step into chaos, anarchy, or (certifiable) insanity; it is a step into true life, meaning, and order. This paradoxical freedom through lessening our grasping for knowledge and control is masterfully celebrated by a fairly recent work on the thought of St. Paul.

In his study on St. Paul, the French philosopher Alain Badiou offers the post-postmodern Western world a blueprint for a truly universal approach to human needs and realities. Central to Badiou's argument is an understanding of Paul's use of the terms *flesh* and *spirit*. For Badiou, these represent the two fundamental options for human existence. The self that is characterized by the flesh is a self who is dead: a self that has accepted a *false consciousness* where it is split off from its own self and potential and is powerless in its universe, since it is *essentially* fragmented, lacking a basic integrity/ability to unify its thoughts and its deeds.[11] A system of thought that reifies this basic gap by increasingly objectifying the self and the world of not-self only intensifies and inscribes this death/despair.

On the other hand, the "spirit" represents an option for the human to move from death to life and is marked by "love"—true regard for others—and an integration of thought (heart, intent) and deed (action). To choose life means to recognize the current state of death and to embrace the power and potential of a new basis for understanding the self and its world; choosing life means to overcome the *essential alienation* of the self with the self and the self with the other (non-self). Since to do so is fundamentally a loss of the mastery or control of the system built on this understanding of self and other, this choice comes at the cost of certainty or knowledge (as characterized by a comprehensive grasp of what is known).

An energy that comes from and functions within this system of alienation (of self with its-self and the non-self) is driven to greater and greater expenditures and demonstrations as a way to stave off the nagging suspicion that this way of approaching the self and the world is a self-destructive pattern. Appetites/"needs" are blown out of proportion as excess, extravagance, and arrogance characterize normal or typical lifestyle expectations. In the West the *excesses* of our energy consumption, consumerism, and entertainment graphically illustrate this scenario. For example, in my geographical context in the oil and gas fields of western Canada, the explosion of this industry gives this situation a particular poignancy and invites a psychoanalytic discussion of denial and overcompensation.

In contrast to this picture of excess, I appeal to a model of the divine as restfully working, being secure in its relation to self and the non-self, and its attitude and appetites being characterized by moderation, restraint, and respect. Within the confidence of its relational fidelity to the Truth of things (a truth that is characterized by the divine person) it can afford to wait, it can afford to fast, it can afford to give its self away. The false consciousness (the flesh/death) is characterized by an approach to energy that sees it as the ability to take, to amass, to wield power and exert control. An authentic consciousness (the spirit/life) grounded in the priority of (and approximate realization of, for the human) personhood sees energy as the ability to give, especially as the ability to give life—to suffer, to die.

By way of analogy, I believe we can discern a quality of existence that shows certain behaviors, behaviors that mimic strength, goodness, love, and so forth, to be inauthentic, to come from a false consciousness. Consider, for instance, a strong man, one who has an accurate assessment of his strength and a corresponding confidence. This individual is the least likely to be out picking fights, responding to every real or perceived slight, or seeking to prove or demonstrate his strength. The truly strong man would most often use his strength, if at all, to do good.[12] This understanding of the nature of God's power—that it does not need to prove itself, nor does it do more than is required, and only then if it will lead to good—is an important feature of this working rest. The rest is in being—confident in its being what it is; the work comes out of the fullness of this being, as is appropriate to that being. This vision, articulated in a variety of ways over the centuries (a theme in Plotinus, appropriated by the mystics, etc.[13]) allows us to see the power of the divine as the ability to give itself away—that is, as love.[14]

We see this divine repose in many ways in the biblical sources: a God who is content with a remnant, a God who chooses the foolishness and scandal of the cross. And we see a consistent human tendency to miss this point: a great prophet who seeks God in the extravagance of Fire and Wind ("but God comes in a still small voice"; I Kings 19); students of the incarnate Son who want to call down fire as judgment (but the Son says, "you do not know what kind of spirit you are of"; Luke 9:54–55). These instances indicate that this misunderstanding of God is attributable, at least in part, to an egocentric (ethnocentric) view and its religious manifestation as a magical/manipulative approach to the divine.[15]

The story of Elijah's confrontation with the priests of Baal is a fascinating representation of this view of the divine (I Kings 18:19ff). His whole approach to the challenge is one that sees energy as external (fire from heaven) and the divine as subject to manipulation. Further, he seeks an external victory, through demonstrable public support, which is realized in part after the decisive answer on the part of Yahweh and the slaying of the priests of Baal. However, this is short-lived, and the royal house (the queen) vows to take his life. As a result, Elijah despairs and God's tender (and more appropriate to the divine) care is given, which culminates in the decisive revelation of God's character mentioned above—God is not in the wind, fire, and earthquake but comes as a whisper. Finally, against Elijah's hope for a widespread victory for God's cause, and his resulting despair, God's encouragement is that He has reserved seven thousand Israelites for himself (seven thousand out of a nation, we can well understand Elijah's despondency).

This egocentric and magical/manipulative approach to reality is the essence of biblical idolatry or false religion, and ancient Jewish religion comes as a direct challenge to such idolatry. Science (and modern technology) in its focus on manipulating the physical universe to yield to the demands and desires of the human is the heir to these magical aspirations.[16] And, as such, science/technology is susceptible to the same prophetic critiques; it too "feeds on ashes; a deceived heart has turned [it] aside . . . [it] cannot deliver [it]self, nor say, 'Is there not a lie in my right hand?'" (Isaiah 44:20) The essential problem is not in the various modes of understanding and harnessing the natural environment (this is essentially a theological task in itself). The problem lies in the inability to determine that that which the human exerts such control over does not warrant human worship! That is, what the human wields control over cannot be the appropriate source for human salvation, by definition of the term.

Instead of providing a real salvation, the basis and practice of current Western culture, as Thomas Berry puts it, resembles the destructive patterns of the alcohol or drug addicted, with all the denial and codependency that these behaviors include. Things are bad, and the destruction is evident, but the continued addictive behavior is the only realistic way to cope in the *short term*. And the psychological/spiritual investment in the behavior still furnishes a form of hope/memory that rationalizes the addiction. On the other hand, change is hard, costly, and contingent. "Any effective cure requires passing through the agonies of withdrawal. If such withdrawal is an exceptional achievement in individuals' lives, we can only guess at the difficulty on the civilizational or global scale."[17]

Berry calls the Western approach to the environment and its environmental and energy crises a mythic addiction. The main thrust of his pioneering article on creative energy is that real change in Western attitudes requires a change at the mythic level. The real issues are not found at the level of current practices or abuses, but in the deep structures that form the basic orientation to the natural world. His call, then, is for an investment in promoting a new mythic structure, one that will lead to a more balanced and sustainable model. I think Berry is correct on two counts: first, that the only real corrective to the current energy crisis will come through a reevaluation of the foundational premises of the materialist worldview of Western culture;[18] second, that the psychic and practical investment in, and commitment to, the current worldview presents a formidable barrier to any real change. Therefore, any realistic option for an alternative approach bears the burden of proof. While not pretending to fully discharge this burden, I think a case can be made for the personalism model in the following ways.

There is widespread testimony to the ideals intrinsic to personalism within the great traditions of human society. For instance, the focus on "being, not doing" as offering the key to complete humanness finds its articulation in ancient Greek and Roman society—the Stoic ideal, for instance. It is also found in the traditions of the East, most notably the concept of wu-wei as elaborated by Taoism and appropriated by Chan/ Zen Buddhism.[19] As an alternative value system with its consequent valuation of energy, there is at least historical evidence/precedent for seeing the personalist model as a viable option. However, these Eastern traditions are primarily focused on "inner" reality, on energy as "being" focused. One can make the case even stronger by looking at the potential for this approach regarding "external" reality, energy as "doing" focused. Here there is also ample evidence within these same traditions of the power of the human mind/soul to be either indifferent to the requirements of the natural world (unsusceptible to pain, hunger, cold, and decay) or to be able to influence/ control natural processes. While these claims to exert control over external processes should not all be taken at face value, there exists at least enough "evidence" to warrant further investigation.[20]

Given the realities of the current dependence on technology and "cheap energy," any alternative model must be able to allow for and manage the current system, even if only as a means of transition. Personalism allows for this and, given the mandate of stewardship of the created order, can be seen to include scientific exploration/development as a central requirement.[21] Yet, as mentioned above, whatever this approach yields, it will not be a commodity subject to patents and packaging. Instead it will be a return to something like the apprenticeship/discipleship models of the ancient world (and still many religious traditions today). This truth, and this power, does not offer itself to all comers and is only earned by an inner transformation and yielding. It remains yet to be seen what an approach to knowledge and truth like this would look like in the midst of the scientific and technological realities of the contemporary West. I can, however, sketch out what I believe would be the formal contours of its unfolding.

Above, I appealed to the "epistemology of the cross" as a way to understand the inverted power structure (from the human standpoint) of divine energy/work. In this section I will appeal to the ethics of the cross, or the radical ethic of Paul, his "ruleless rule," as outlined in I Corinthians: (1) "All things are lawful," *but* (2) "not all are profitable," (3) "not all things edify," and (4) "I will not be brought under the power of any"(I Corinthians 6:12; 10:23).

First, "all things are lawful": I would see this as requiring that a model of energy would be open to new ideas, directions, and uses. Presumably this would include some form of decentralized control over ideas/concepts/approaches, especially freeing energy from some of the more insidious forms of capitalist enterprises—where market shares are monopolized and new ventures are decided upon by capital gains projected for the current stakeholders in the existing market.

Second, "not all things are profitable": this suggests that the model would look at the total impact (short-term, long-term, systemic) on the proposed energy developments.[22] What would maximize the overall benefit for all affected parties? Since the premise of this energy model is personalism, profit/benefit cannot be decided by looking at economics or human needs/interests alone. Ecological issues and sustainability must be considered important factors in the decision. The golden rule, "treat others in the manner you want them to treat you"—the ethical aspiration of both utilitarianism and Kant's duty ethic—on the basis of personalism, extends to the whole natural order; although, as distinct to pantheism (or forms of spiritism) these are not seen as personal or as the person herself, but as having value by virtue of having an intrinsic relationship to, or being an expression of, that person.

Third, "not all things edify": this makes the well-being, wholeness, and growth of the self and others (the poor, marginalized, dispossessed) a goal for energy development. The key here is a focus on the development of human potential toward a flourishing life.[23] Given the premise of this system, this does not mean more "stuff," or even more "opportunities," although, given the rationale behind something like Maslow's hierarchy of needs, it would require meeting basic survival needs as a minimum. It would not, however, content itself with this, but would seek further development along the lines of spiritual and psychological well-being, (like Maslow's "actualized" person). The management of physical energy resources would be guided by questions like: Does this system restrict the benefits of energy to those who can afford it? Does the system allow for (either intrinsically or extrinsically) an overall net gain for human culture as a whole? How are economic profits utilized?

Fourth, "I will not be brought under the power of any": this criterion is the most abstract or nebulous. In the spirit of this whole enterprise of a "ruleless rule" for energy development, we realize that we cannot satisfy ourselves with categorical pronouncements. Enslavement to our wants, needs, desires is often a subjective or personal thing and thus requires some honest and candid self-assessment on the part of the leaders/developers in the energy field and culture as a whole. For instance, the use of fossil fuels for current industry and lifestyle in the West may be allowable ("all things are lawful"), but more consideration should have been given to the very real danger that our complete dependence upon these energy sources poses at *some time*. Other dependency issues arise in the use of pharmaceuticals and other solutions as responses to the energy needs of human physiology, psychology, and spirituality. These approaches reinscribe the basic error of a materialist approach by acting on the external, treating symptoms and ignoring the holistic or systemic needs and issues. As forms of energy they are agents that enable, not empower.

I see these four briefly stated concepts as forming some general principles that could aid in responding to the crisis situation that a reliance on "cheap" energy has created. These principles would also help manage the existing cultural and economic structure that is based on these sources of energy. In the long run, espousing such a value system would work to transform the deep structure of Western consciousness so that a shift in the priorities that direct energy would be feasible. I suggest that this new worldview would be marked by a move away from the external manifestations of power/energy and a move toward the poise and repose that are characteristic of the true, enduring source of power/energy.

I have proposed a radical shift in our understanding of energy and our consequent uses of energy. This is guided by the value of *well*-being. In the *Nicomachean Ethics* Aristotle claims that in order to decide if human life has been lived well (the good human), one first needs to know what the human is meant or intended for—the same way the judgment that someone is a good guitar player requires knowing what a guitar is made for. This claim demands some consideration. As a culture, the West needs to ask the question "Why?" Why would we want to solve the "energy crisis"? What is all this busy-ness and energy expenditure about? What purpose does it serve? Does it meet the Big Purpose? Using a theological understanding of Being as person, as the basic structure of reality, I propose that realizing the life-giving, life-enhancing potential of the human person—becoming a complete person—is the best response to Aristotle's challenge. And, further, if this is the goal, then critical/essential energy is that which moves us toward this goal.

CHAPTER 10

# God Is Green; or, A New Theology of Indulgence

*Jeffrey W. Robbins*

We all contribute to climate change, but none of us can
individually be blamed for it. So we walk around with a free-
floating sense of guilt that's unlikely to be lifted by the purchase
of wind-power credits or halogen bulbs. Annina Rüst, a Swiss-
born artist-inventor, wanted to help relieve these anxieties by
giving people a tangible reminder of their own energy use,
as well as an outlet for the feelings of complicity, shame and
powerlessness that surround the question of global warming.

So she built a translucent leg band that keeps track of your
electricity consumption. When it detects, via a special power
monitor, that electric current levels have exceeded a certain
threshold, the wireless device slowly drives six stainless-steel
thorns into the flesh of your leg. "It's therapy for environmental
guilt," says Rüst, who modeled her "personal techno-garter"
on the spiked bands worn as a means of self-mortification
by a monk in Dan Brown's novel "The Da Vinci Code."

NEW YORK TIMES MAGAZINE, 8th Annual Year in Ideas

All those who believe themselves certain of their own
salvation by means of letters of indulgence, will be
eternally damned, together with their teachers.

MARTIN LUTHER, THESIS 32, from *The Ninety-Five Theses*

In 1940, shortly after speaking with Colonel Oster at the *Abwehr* meeting
during which he was enlisted in the plot against Adolf Hitler that would
eventually lead to his imprisonment and execution, Dietrich Bonhoeffer
penned what remains one of the most scathing attacks against theoretical
or systematic ethics. As he wrote in the opening paragraph of the chapter
"Ethics as Formation," ethical reasoning had become entirely superfluous,
not because of indifference and certainly not because of irrelevance. "On

the contrary," Bonhoeffer wrote, "it arises from the fact that our period, more than any earlier period in the history of the west, is oppressed by a superabounding reality of concrete ethical problems."[1] In addition to the failure of moral theorists, Bonhoeffer also exposes the ethical failings of so-called *reasonable* people, ethical purists—or in his words, *fanatics*—who believe they "can oppose the power of evil with the purity of [their] will and of [their principle]," the solitary individual of *conscience*, who "fights a lonely battle against the overwhelming forces of inescapable situations which demand decisions"; the person of *duty*, who "will end by having to fulfill his obligation even to the devil; and the privately *virtuous*, who "knows how to remain punctiliously within the permitted bounds which preserve him from involvement in conflict" but who thereby "must be blind and deaf to the wrongs which surround him."[2] Bonhoeffer saw each of these as false, or at least insufficient, paths to meet the challenge of his age, and in what is perhaps the most recognized sentence from this fragmentary and unfinished work, he issues a clarion call: "Yet our business now is to replace our rusty swords with sharp ones."[3]

Here we stand yet again at such a moment with the global economy in tatters, the U.S.-led war on terror ongoing, an energy market that is wildly fluctuating, and the global demand for energy escalating at an unsustainable pace that will only compound the environmental degradation already wrought. We stand intimately interconnected and interdependent with those around the world by the apparent triumph of global capital, but with outmoded political institutions and ideas to rein in what the political theorist Benjamin Barber rightly identifies as "savage capitalism."[4] And our religion, long caught up either in the culture wars and the micro-politics of identity or voluntarily confining itself as little more than self-help therapy, finds itself once again thrust in the public domain forcing a fundamental theo-political reevaluation of the basic modern liberal assumptions of secular sphere as the naked public square.

In what follows, I offer three snapshots from a generation past. My purpose is not, unfortunately, to suggest the way forward but, more modestly, simply to raise the stakes of our present discussion on theology and energy. The questions asked by this book, while urgent and timely, are not new. They are the very questions and issues with which we have been faced for some time now, the very choices we have been asked to make but have long deferred, the very crisis seen on the horizon and predicted by select lone voices that is now rapidly approaching. Perhaps now, with all of us a bit more knowingly feeling the oppression of the "superabounding reality of concrete ethical problems" we face, we stand poised to act and think otherwise.

## I

In 1973, largely inspired by and in response to the debacle at the Bureau of Indian Affairs in 1972 when a group of Indian activists captured and destroyed parts of the federal building in Washington, D.C., Vine Deloria Jr. published the book that would become one of the defining works in the American Indian movement, *God Is Red: A Native View of Religion*.[5] In this book, Deloria continues where he left off in his "Indian Manifesto" from 1969, *Custer Died for Your Sins*, by chronicling the cultural, political, and religious impasse that continues to exist between Indians and non-Indians in the United States. As he wrote in *God Is Red*, "The impasse seems to be constant. Indians are unable to get non-Indians to accept them as contemporary beings" (56). Thus the significance of the Native Americans' violent protest in 1972: "Indians were no longer the silent peaceful individuals who refused to take dramatic steps to symbolize their grievances" (6). Instead, Indian activists claimed their place on the stage of the larger civil rights movement. But in borrowing from the playbook of other increasingly radicalized militant civil rights groups from the late 1960s and early 1970s, the Indian protesters "had become simply another protest group" (6), which confirmed for Deloria yet again the "peculiar tragedy" not only of the Indian movement but also of the much broader history of relations between Indians and non-Indians throughout U.S. history—namely, Native Americans have "never been able to influence the intellectual concepts by which Americans view the world" (38).

For Deloria, this irony of the Indian movement was palpable. By expressing its grievances in terms of the politics of identity characteristic of the civil rights era, it was unclear for Deloria what to make of the Indian movement. Was it simply the final spasm of the revolutionary 1960s from a largely overlooked and forgotten minority group? By their violent confrontation with the federal government, was this a tactical advance in a broader social and political strategy? Or was it simply a sign of frustration and despair? After all, as Deloria asserts, "Few Indians ever accepted the premises of the Civil Rights movement" (59), because it remained within the parameters of an essentially Euro-American Christian worldview.[6] Further still, when considering the Indian movement in the broader context not only of civil rights but also together with the anti–Vietnam War protests, the rise of the counterculture, and the beginnings of the environmentalist movement, according to Deloria's analysis, "they can all be understood as desperate efforts of groups of people to flee the abstract and find authenticity, wherever it could be found" (64). And therein lies the irony, because it is at precisely this point when the Western world seemed to be either caught up in its own violent upheaval or at the point of its own self-exhaustion that "Indians [became] popular" (64)—but only, so Deloria argues, by becoming something that they were not.

The one sure sign of their sudden popularity was the money that flowed in from the mainline Protestant denominations. As Deloria writes, progressive-minded Protestants "were ecstatic when informed by Indians that they were guilty of America's sins against the Indians." By expressing their solidarity with the Native Americans, these well-intentioned WASPs could effectively "purchase indulgences for their sins by funding the Indian activists to do whatever they felt necessary to correct the situation" (58–59). While Deloria himself was disgusted by the spectacle of once-proud Indian activists becoming "little more than puppets dancing for liberal dollars" (65), he was equally harsh in his assessment of the Christian churches who "bought and paid for the Indian movement and its climactic destruction of the Bureau of Indian Affairs" (59).

After all, the road to hell is paved with good intentions, and while the buying of indulgences might soothe the conscience of the damned, it does little to nothing in altering the intellectual categories or concepts by which the world is viewed, which, as we have already seen, was Deloria's prime concern. Too often, however, rather than being the cause for the dramatic and fundamental shift in the intellectual conception of the world, the American turn toward the native was in fact an extension of the right to ownership and possession that had driven the American Indians from the land in the first place. For instance, in the chapter derisively titled "America Loves Indians . . . And All That," Deloria recounts the repeated incidents of white Americans looting Indian remains, collecting them as artifacts, and somehow thinking this act was an expression of the highest form of respect. The prerogative of American manifest destiny by which non-Indians took possession of the country's land takes the form of a plundering of the past as a misguided quest for authenticity, the American Indian "seemed to hold the key to survival" (64). America looked to its natives in order to save it from itself, but instead of allowing the native view of religion to call to question America's basic orientation to reality, it instead, once again, co-opted the natives on its own terms. This time, however, it was not the land, but the very conception of land that was at stake.

Indulgent and misguided though the appropriation of the Native American was, for Deloria it nevertheless posed a fundamental choice. Whereas American Christians remained mired in their collective guilt, the native view of religion claimed by Deloria was oriented around ecology. As he writes, "the choice appears to be between conceiving of land as either a subject or an object" (70). American Christianity "has avoided any religious consideration of ecological factors in favor of continuous efforts to realize the Kingdom of God on earth" (72). And whether progressive or reactionary, liberal or conservative, when it comes to our religious understanding of the land, "neither left-wing nor right-wing Christianity appears to understand the nature of the ecological disaster facing us. Rather they both seem to vest their faith in the miraculous ability of science to solve the problem of the dissipation of limited resources" (72). This concern with ecology, which is fundamental to the native religious conception of the world, requires more than "the relatively simple admission of guilt before ecological gurus" (74); it involves a fundamental reorientation regarding the very locus of meaning. In short, it is not forgiveness that is needed but restitution—restitution to the rightful "spiritual owners of the land" whereby the land is reconceptualized as inhabited—indeed, as living—space. And, more direct for our purposes, it raises skepticism with regard to those who would place too much faith in science and technology to solve the ecological disaster—and by extension, the energy crisis—that we currently face as yet another misguided effort to realize the Kingdom of God on earth.

Thus, in contrast to the death-of-God theologies that made their appearance in the United States a decade earlier, this sudden embrace of the Indian as the living alternative to all things Western led to Deloria's alternative proclamation that "God Is Red." As Deloria tells it, "Perhaps we have come to realize that Western man cannot find his way in society either by demythologizing his condition as Kierkegaard, Nietzsche, and the social gospel people have attempted or remythologizing it as Billy Graham and the Fundamentalists have tried to do. Perhaps an entirely new analysis of the nature of society must be undertaken, perhaps a new understanding of the nature of religion must be found" (68–69).

From Deloria we get an argument for a native approach to religion that would bring matters of ecology, land, and space to the fore. Something other, more fundamental than the cycle of sin, penance, and redemption was required—something more radical even than the seizing of a federal building in militant protest. Deloria goes on to make the case that this conceptual revolution would actually show religion to be more compatible with contemporary science and technology. When God is Red, so the argument goes, the greening of religion is sure to follow. This approach has yielded much positive benefit even within more traditional theological circles by those who are increasingly concerned with, and define themselves in terms of, eco-theology.[7] What is still needed, however, in addition, but in no way in opposition, to these expressed concerns is a focus on energy. The chance remains, once again or perhaps for the first time, to finally, even if far too late, recognize the Indian as our contemporary.

## II

In 1973, armed members of the American Indian movement and the Lakota Nation made a final stand for native rights in a siege of the town of Wounded Knee in South Dakota in a conflict that lasted for seventy-one days and eventually led to the arrest of twelve hundred people and effectively signaled the end of the Indian movement.

Meanwhile, the Yom Kippur War began and ended, lasting a total of only three weeks, but having far-reaching implications that extend to this present day, including the decision by OPEC to restrict the flow of oil to all countries allied with Israel, which resulted in the immediate spiking of oil prices by 200 percent.

The fight over a promised land consolidated the power and influence of the oil-producing states in the Middle East and beyond as it was realized that oil is more than simply energy and a commodity. The weaponization of oil left all matters of energy inextricably entangled in discussions of power, money, and politics.

In reaction to the Arab Oil Embargo of 1973, the United States developed the Strategic Petroleum Reserve (SPR) to provide an emergency stockpile of an approximately two months' national supply of oil. This hoarding of energy, which has now long substituted for a coherent national energy policy, was meant to insulate the United States from global petropolitics and to stabilize the oil market. But as Lisa Margonelli, author of *Oil on the Brain*, describes it, it is now "serving double duty as a defense and a target," further politicizing oil as a global commodity and perhaps even artificially inflating the price of oil by as much as 25 percent. "In a sense," Margonelli writes, "the SPR is a monument to the cataclysmic oil crisis of 1973. Like all monuments, it documents both the size of the shock and the sincerity of the emotion—700 million barrels! It's also become a bit of nostalgia, out of step with the world around it." Put succinctly, "The SPR was cold war thinking translated to oil."[8]

As a monument to the long-term impact of 1973, the SPR signified a last-grasp effort by a nation threatened with a crisis of identity. The same cultural malaise and sense of exhaustion described contemporaneously by Deloria has been captured by Margonelli as well. Americans had long come to define themselves by the sense of open space and endless expanse. The open road was not only the image of the beat generation but also the promise of an America structured by its sense of time always in the future. America's love affair with its cars, therefore, represents its perpetual capacity at reinvention, the potential to pick up and move somewhere different, someplace still to be discovered and yet to be tamed. "The whole definition of being American," Margonelli writes, "was that we drove our cars anywhere we wanted to." But "with the oil crisis, the long upward expansion of the U.S. economy since World War II ended and started to reverse." The "oil shortage brought about a deep psychological insecurity" that, in the words of the U.S. comptroller general, gave Americans the "illusion of U.S. impotence" (104).

In this sense, and in a remarkably similar way to how the appropriation of the American Indian was a purchase at redemption for a nation exhausted by its own self-indulgence, much more than any economic or political impact it has had, the SPR proved to be a psychological cure to what was perceived to be merely a psychological problem. After 1973, Americans needed to get their swagger back. The investment in the SPR was a down payment for a generation's indulgence. For an illustration, compare and contrast the difference between Travis Bickle in Martin Scorsese's 1976 masterpiece *Taxi Driver* and Sylvester Stallone's rejoinder from 1982 in the first incarnation of the Rambo series, titled *First Blood*. Both Travis Bickle and John Rambo were Vietnam veterans who could not quite manage to assimilate themselves back into mainstream American society, both carried with them the American surrender in Vietnam as a personal trauma, and both externalized that internal self-loathing in a climactic act of violence. The difference is that by Scorcese's 1976 rendering, Travis Bickle is an antihero, as his bloodlust actually reveals his own death wish. Whereas when Rambo is given the chance to make things right, this film from 1982 actually follows the script of the myth of redemptive violence, thus restoring the American soldier to his proper heroic archetype. Or contrast Jimmy Carter's sweater vest to Ronald Reagan's "morning in America." In each case, the reality of the United States' vulnerability is replaced by a fiction and the American dream is restored. Returning to Margonelli, she concludes her analysis of the SPR by writing that it "offers the illusion of safety but no real insurance. It is too small, too centralized in the Gulf Coast, and too vulnerable to be effective. Insulating the United States against oil shocks requires more work than just stockpiling" (115).

When it comes to America's energy policy, which is to say nothing of its view or fundamental conception of energy, we need much more than psychological reassurance. When we consider the long journey oil makes from around the world to fuel our cars, to heat our homes, and to be consumed as commercial products, we realize that it is much more than a financial commodity. The United States—indeed, the entire world in the age of globalization—enjoys an entire culture of oil. Better still, as Kevin Phillips has shown in his book *American Theocracy*, the ascendancy of the United States as the world's lone superpower is predicated entirely on cheap and readily accessible oil.[9] In short, the Age of Oil is the Age of American Supremacy, and as Phillips adds, "Oil abundance has always been part of what America fights *for*, as well as *with*" (3; emphasis in original). But with this great strength also comes great vulnerability: "The politics of oil dependency in the U.S. is ingrained and possessive—a culture of red, white, and blue assumptions of entitlement, a foreign policy steeped in covert petroleum emphasis, and a machismo philosophy of invade-and-take-it" (33), which leads Phillips to describe the United States as a "vulnerable oil hegemon" (12), likening the country to an imperial power whose best days are behind it, much like the Dutch when wind and water were the primary sources of energy and expansion and the British in the age of coal.

Thus, the persistence of the U.S. culture of oil requires the passivity of the American citizen as consumer. Regardless of the posted price of a gallon of gas at the local station, there is a high price for oil that implicates the entirety of American foreign policy, the stranglehold of the military industrial complex, and our own peculiar brand of empire. From the depleted "black giant" of the Texas plains; to the geopolitics of petro-states such as Iran, Venezuela, and Russia; to the maneuvering of a Nigerian warlord and the rapidly expanding oil markets of China and India, our dependence on oil has global repercussions that are economic to be sure, but just as assuredly moral and religious. As Paul Roberts has written, "Our brilliant energy success comes at great cost—air pollution and toxic waste sites, blackouts and price spikes, fraud and corruption, and even war. The industrial-strength confidence that was a by-product of our global energy economy for most of the twentieth century has slowly been replaced by anxiety."[10]

What to do, then? Margonelli's recommendation is for a cultural revolution by which we rethink our entire relation to energy, reconceptualize space, and develop an ethic of conservation and efficiency rather than consumption.[11] What is also needed, and this is where Phillips's analysis serves as a critical supplement, is a theo-political critique of the religiosity that is at the basis of America's culture of consumption. In sum, as Americans "cling to and defend an ingrained fuel habit" (54), we have been immersed in at least a Thirty Years' War over Middle Eastern oil. To explain this incredible shortsightedness, Phillips chronicles the political rise of the religious right and the broad-based cultural fascination with apocalypticism: "To sketch the revival-prone sectarian and radical side of American religion: Its increasing presence is breeding a politics of cultural narrowness, moral and biblical bickering, revivalism in the White House, and international warfare to spread the gospel, fulfill the Book of Revelation, or both" (121). While this American parochialism and fascination with the end times might not be new, Phillips argues that it "take[s] on much greater importance now as Christian, Jewish, and Muslim holy lands occupy center stage in world politics and as sites of military confrontation" (103). Expanding the analysis even further, Phillips outlines the triple threat we currently face:

1. The increasing domination of U.S. policy by the hunger for cheap oil in a world of dwindling supplies, which has led in turn to an obsession with projecting U.S. power across the endlessly volatile Middle East.
2. The Republican Party seriously under the sway of Christians who believe in biblical inerrancy and a reading of Scripture that inspires them to apocalyptic obsessions with that same part of the world.
3. The headlong growth of American debt of all kinds—household spending, a massive trade gap and a federal deficit that leaves American policy susceptible to the foreigners who buy the securities that keep the U.S. government afloat, and who could sink it with the decision to stop buying.[12]

## III

Returning for one more snapshot from the early 1970s: In 1971, in a policy measure that has come to be known as the "Nixon Shock," President Richard Nixon unilaterally withdrew the United States from the Bretton Woods agreement and stopped the direct convertibility of the U.S. dollar to gold. The Bretton Woods agreement had established the International Monetary Fund and the World Bank and had reestablished a modified version of the gold standard in an effort to bring stability to the global economy in the wake of World War II. Nixon felt forced into this decision to withdraw from the agreement by the realities of accelerated inflation caused by an escalating U.S. trade deficit. Though the policy "closed the gold window" by making the U.S. dollar inconvertible to gold, Nixon nevertheless tried to maintain a fixed exchange rate. But this also quickly proved to be unfeasible, and by 1976 all of the world's major currencies were allowed to float, thus giving birth to the new global economic condition of high finance.

Phillips's articulation from *American Theocracy* of the triple threat the United States currently faces makes clear the integral connection between energy, religion, and money. Likewise, Mark C. Taylor sees a broad cultural significance to the so-called Nixon Shock. As Taylor writes in *Confidence Games*, "It is no exaggeration to insist that going off the gold standard was the economic equivalent of the death of God. God functions in religious systems like gold functions in economic systems: God and gold are believed to be firm foundations that provide a secure anchor for religious, moral, and economic values. When this foundation disappears, meaning and value become unmoored and once trustworthy symbols and signs float freely in turbulent currents that are constantly shifting."[13] One irony, as pointed out on the editorial pages of the *Wall Street Journal* from January 2008, is that while the U.S. currency has been allowed to float, the price of oil and gold have run in almost perfect tandem since 2001.[14] For some this suggests that oil has become the new gold standard, replacing the American dollar as the true measure of value. If so, it conforms to Taylor's analysis, only with an added twist. If gold once functioned as a firm foundation by which to secure economic value, oil is a slippery and volatile sludge. While the value of gold can be traced to its usefulness as coinage— thus, as the necessary third that makes possible the transformation from a basic barter economy—oil's value is in its convertibility to various petro-products. As such, oil is not only the fuel for the global economy but its product of exchange as well.

If the collapse of the gold standard is the equivalent to the death of God, then oil as the new gold standard can be likened to the postmodern return of religion and suggests a world not only where meaning and value have become unmoored but, even more, one of infinite complexity and codependence wherein the ecological, theological, and economic realities are all deeply intertwined: a world with oil and gas pipelines crossing international borders where the international waters of the oil tankers' trade routes are policed almost exclusively by the American military. Meanwhile, the American government is effectively impotent at controlling, or even safely predicting, the price of a gallon of gasoline. It is a world where business interests force cultures to collide, and where religions' claim for sacred space inevitably runs up against global capital's demand for open borders and the privatization of land. It is a world wherein the theological preoccupation with one's own private relation with God and the repeated cycle of sin and atonement no longer suffices as an adequate religious orientation for our time. Taylor describes it as a "world without redemption." We might just as well call it a world at war, the inescapable consequence of the pegging of all value to that which is explosively volatile, that which is toxic, and that which is at or near its peak, and thereby that which is bound to disappoint.

In his book *Religion and Capitalism* British philosopher of religion Philip Goodchild asks the question: What is the price of piety, especially considering how money has become our new God? In answer to this question, I offer the image of the oil gusher as a representation for the postmodern return of religion. An oil gusher, which has been cinematically portrayed in such movie classics as *Giant* and more recently in *There Will Be Blood*, is an uncapped oil well connected to a deep reservoir of oil under intense pressure. When unleashed, the oil can shoot two hundred feet or higher into the air and includes not only the oil but the sand, mud, rocks, and water as well. On the one hand, it might be considered an image of plenty, of sudden and ever-renewing wealth. But in point of fact, an oil gusher is to be avoided at all costs, an image of waste rather than wealth, a dangerous explosion that hardened oil wildcatters know has the power to kill. It can be, and is, effectively managed by current subterranean drilling techniques, but stands as a visual reminder of the volatility—indeed, the very explosive potential—of a global consumer economy bankrolled by cheap oil.

The perils of the petro-states on geopolitics are well documented. Around the globe, there is seemingly a direct relation to vast mineral resources and political corruption and the stratification of society. With the easy money associated with oil there is little demand for transparency or accountability from the people. With the vast transfer of wealth from the oil-importing states to the oil-exporting states has come the inordinate influence of Iran and Saudi Arabia in fueling radical Islam. Also, as Phillips makes clear, it has led to the inordinate influence of southern evangelicals in American politics. Perhaps we have done everything in our power to keep this volatile mix in check, or safely beneath the surface. Perhaps at least a half-century's worth of realpolitik has provided for the ready supply of oil our way of life demands. Perhaps we might continue to pay down the price of these indulgences so that our way of life and very mind-set might remain undisturbed, unsullied by the despoiled and tarred landscape we have left behind. Or perhaps instead the gusher is about to blow in what is bound to be a wasteful and destructive orgy of violence.

There is only so long the well can remain tapped, only so long before our endless consumption consumes us, only so long before our purchase of indulgences comes due at the day of reckoning. If God is Green, the choice still remains, whether this be a token gesture toward our collective ecological guilt, a realization of how money stands in for God after the death of God, or indeed the spiritual and theological reformation long sought.

NOTES

INTRODUCTION
*Donna Bowman and Clayton Crockett*

1. Flora A. Keshgegian, *God Reflected: Metaphors for Life* (Minneapolis: Fortress Press, 2009), 154.

2. Iain Nicholson, *Dark Side of the Universe: Dark Matter, Dark Energy, and the Fate of the Cosmos* (Baltimore: Johns Hopkins University Press, 2007), 139.

1. THE ENERGY WE ARE: A MEDITATION IN SEVEN PULSATIONS
*Catherine Keller*

1. See Glen A. Mazis, *Earthbodies: Rediscovering Our Planetary Senses* (Albany: State University of New York Press, 2002).

2. William Blake, *The Complete Poetry and Prose of William Blake* (New York: Anchor Books, 1965), 34.

3. Ibid.

4. See William Blake, "To Nobodaddy," in *The Complete Poetry and Prose of William Blake*, ed. David V. Erdman and Harold Bloom (1962; New York: Anchor Books, 1985), 471.

5. Robert Pollin, "Doing the Recovery Right," *Nation* 288, no. 6 (2009): 13–18.

6. Ibid., 18.

7. George Monbiot, "One Shot Left," *Guardian* (25 November 2008). Available at http://www.monbiot.com/archives/2008/11/25/one-shot-left. Accessed 21 December 2009.

8. Katey Walter Anthony, "Methane: A Menace Surfaces," *Scientific American* (Dec. 2009): 9.

9. Richard Hawkins, Christian Hunt, Tim Holmes, and Tim Helweg-Larsen, *Climate Safety* (United Kingdom: Public Interest Research Centre, 2008), 2.

10. Alfred North Whitehead, *Process and Reality: An Essay in Cosmology* (1978; New York: Free Press, 1985), 167.

11. Ibid., 309; emphasis mine.

12. Jay McDaniel, *Gandhi's Hope: Learning from World Religions As a Path to Peace* (Maryknoll, N.Y.: Orbis Books, 2005), 38.

13. Roland Faber, "Ecotheology, Ecoprocess, and Ecotheosis: A Theopoetical Intervention," *Salzburger Zeitschrift für Theologie* 12 (2008): 75–115. Available at http://faber.whiteheadresearch.org/files/FaberR-Ecotheology_Ecoprocess_and_Ecotheosis.pdf, 99. Accessed 4 January 2010.

14. Ibid., 115; emphasis in original.

15. Ibid.

16. See my own brief theological appropriations of chaos and complexity theory in Keller, *Face of the Deep: A Theology of Becoming* (London: Routledge, 2003, ch. 11); also a more introductory version in Keller, *On the Mystery: Discerning Divinity in Process* (Minneapolis: Fortress, 2008). See also Ilya Prigogine and Isabelle Stengers, *Order Out of Chaos: Man's New Dialogue with Nature* (Boulder, Col.: New Science Library, 1984), in which Whitehead plays a significant role.

17. Bruno Latour, *Politics of Nature: How to Bring the Sciences into Democracy*, trans. Catherine Porter (Cambridge, Mass: Harvard University Press, 2004), 5; emphasis in original.

18. Whitehead, *Process and Reality*, 339.

19. Stephen Moore and Mayra Rivera, eds., *Planetary Loves*. Fourth Drew Transdisciplinary Theological Colloquium (New York: Fordham University Press, forthcoming).

20. John B. Cobb, *The Earthest Challenge to Economism: A Theological Critique of the World Bank* (New York: Macmillan, 1999).

21. Whitehead, *Process and Reality*, 309.

22. Ibid.

23. Ibid., 116.

24. Shimon Malin, *Nature Loves to Hide: Quantum Physics and Reality, a Western Perspective* (New York: Oxford University Press, 2001), 163.

25. Ibid., 186.

26. For accessible introductions to the concept of quantum entanglement, see Brian Greene, *The Fabric of the Cosmos: Space, Time, and the Texture of Reality* (New York: A. A. Knopf, 2004); Louisa Gilder, *The Age of Entanglement: When Quantum Physics Was Reborn* (New York: Alfred A. Knopf, 2008); Anton Zeilinger, *Dance of the Photons: From Einstein to Quantum Teleportation* (New York: Farrar, Straus and Giroux, 2010).

27. Malin, *Nature Loves to Hide*, 186.

28. Bernard D' Espagnat, *On Physics and Philosophy* (Princeton, N.J.: Princeton University Press, 2006); David Bohm, *Wholeness and the Implicate Order* (1980; New York: Routledge, 2002).

29. First made vivid for me, with these examples, by Bill McKibben's prophetic *The End of Nature* (New York: Random House, 2006).

30. Lee Smolin, *The Trouble with Physics: The Rise of String Theory, the Fall of a Science* (New York: Houghton Mifflin Harcourt, 2007), 15.

31. Percy Seymour, *Dark Matters: Unifying Matter, Dark Matter, Dark Energy and the Universal Grid* (Franklin Lakes, N.J.: Career Press and New Page Books), 193.

32. Alfred North Whitehead, *Science and the Modern World* (New York: Free Press, 1925), 91.

33. Seymour, *Dark Matters*, 194.

34. The brilliant or luminous darkness refers in the tradition of negative or apophatic theology, especially associated with Gregory of Nyssa (fourth century) and Denys (sixth century), to the unknowable depth of divinity. See Chris Boesel and Catherine Keller, eds., *Apophatic Bodies: Negative Theology, Incarnation, and Relationality* (New York: Fordham University Press, 2009).

35. Sallie McFague, *The Body of God: An Ecological Theology* (Minneapolis: Fortress Press, 1993). McFague was systematically amplifying the metaphor of process philosopher Charles Hartshorne, *Omnipotence and Other Theological Mistakes* (Albany: State University of New York Press, 1984).

36. Job 38:8.

37. Keller, *Face of the Deep*.

## 2. THE FIRE EACH TIME: DARK ENERGY AND THE BREATH OF CREATION
### *Mary-Jane Rubenstein*

1. I am indebted to Clayton Crockett for the invitation to think about energy in the first place, and to Aristotle Papanikolaou, whose thoughts on Palamas and Lossky got this project off the ground. My thanks also to Brian Wecht at the Institute for Advanced Study, whose revisions to the scientific material have been invaluable. Brendan Conuel read an early draft, and I am grateful for his suggestions and reflections. While these interlocutors may be credited for much of my understanding of these sources, none of them bears any responsibility for what I have done with (or to) them.

2. Albert Einstein, "Cosmological Considerations on the General Theory of Relativity (1917)," in *Cosmological Constants: Papers in Modern Cosmology*, ed. Jeremy Bernstein and Gerald Feinberg (New York: Columbia University Press, 1986), 16–26.

3. Dan Hooper, *Dark Cosmos: In Search of Our Universe's Missing Mass and Energy* (New York: Smithsonian Books, HarperCollins, 2006), 143. For the next sixty years, the cosmological constant became "theoretical poison ivy"; it was just assumed that lambda had a value of 0. It was not until the late 1990s that observational physicists began detecting a lambda value that

was not 0; in other words, Einstein may have thrown out the cosmological constant too soon. Dark energy might be it. See Robert P. Kirshner, *The Extravagant Universe: Exploding Stars, Dark Energy, and the Accelerating Cosmos* (Princeton, N.J.: Princeton University Press, 2002), xi, 215–21.

4. A. Deprit, "Monsignor Georges Lemaître," in *The Big Bang and Georges Lemaître*, ed. A. L. Berger (New York: Springer, 1984), 370. Lemaître's findings were initially published as Georges Lemaître, "Un Univers Homogène de Masse Constante et de Rayon Croissant Rendant Compte de la Vitesse Radiale des Nébuleuses Extragalactique," *Annales de la Société Scientifique de Bruxelles* 47 (April 1927). A similar formulation had been posited two years earlier by a Russian mathematician and meteorologist named Alexander Friedmann (see Brian Greene, *The Fabric of the Cosmos: Space, Time, and the Texture of Reality* [New York: Vintage, 2005], 230).

5. Note that the French atome bears as much resemblance as its English counterpart to "Adam." See Georges Lemaître, "The Expanding Universe," *Monthly Notices of the Royal Astronomical Society* 91 (1931): 490–501. Of course, the word atom originally comes from the Greek *atomos*, meaning "uncuttable." The notion that the world is composed of such elementary units was set forth by Leucippus and his pupil Democritus in the fifth and fourth centuries c.e. Atomism was roundly opposed by Plato and Aristotle and then revived more or less for good by the English chemist and physicist John Dalton (1766–1844). Niels Bohr's quantum mechanical theory of the atom was in place by 1911, well before Lemaître formulated his hypothesis.

6. Hooper, *Dark Cosmos*, 144.

7. "When God began to create heaven and earth—the earth being unformed and void, with darkness over the surface of the deep (*tehom*) and a wind from God sweeping over the water" (Genesis 1:1–2; JPS translation). See Catherine Keller, *Face of the Deep: A Theology of Becoming* (New York: Routledge, 2003).

8. "Since Hubble, cosmologists had been trying to measure the slowing of the expansion due to gravity; so expected was slow-down that the parameter used to quantify the second derivative of the expansion, $q_o$, was called the deceleration parameter" (Joshua A. Frieman, Michael S. Turner, and Dragan Huterer, "Dark Energy and the Accelerating Universe," *arXiv* (March 7, 2008).

9. For a surprisingly thrilling account of these two teams' concurrent efforts, see Kirshner, *Extravagant Universe*, esp. 158–232.

10. See Sean M. Carroll, "Why Is the Universe Accelerating?" in *Carnegie Observatories Astrophysics Series, Vol 2: Measuring and Modeling the Universe*, ed. W. L. Freedman (Cambridge: Cambridge University Press, 2003). See also Dennis Overbye, *Lonely Hearts of the Cosmos: The Story of the*

*Scientific Quest for the Secret of the Universe* (New York: Back Bay Books, 1999), 428–436.

11. Paul J. Steinhardt and Neil Turok, *Endless Universe: Beyond the Big Bang* (New York: Doubleday, 2007), 43–44. See NASA's summary of the data WMAP has gathered at http://map.gsfc.nasa.gov.

12. Hooper, *Dark Cosmos*, 138.

13. Turner and his colleagues have also considered the possibility that the universe's expansion indicates a breakdown of General Relativity at cosmological scales rather than the presence of a counter-gravitational force. See Frieman, Turner, and Huterer, "Dark Energy and the Accelerating Universe."

14. "Apophatic" theology is literally a theology of un-saying—a strategy of reaching the limits of language in order to transcend them, becoming united to the God who exceeds all human determinations (including the distinctions of light and darkness, presence and absence, and nearness and farness). "Pseudo-Dionysius" was most likely a sixth-century monk writing in present-day Syria, who wrote in the name of Dionysius the Areopagite, Paul's Athenian convert named in Acts 17:34. Because of his presumed proximity to St. Paul, Dionysius's works were taken by the medieval church to have near-apostolic authority. For the history of his reception, see Sarah Coakley and Charles M. Stang, eds., *Re-Thinking Dionysius the Areopagite: Directions in Modern Theology* (London: Wiley-Blackwell, 2009).

15. Hooper, *Dark Cosmos*, ix.; and Turner cited in Ron Cowen, "Embracing the Dark Side: Looking Back on a Decade of Cosmic Acceleration," *Science News*, February 2, 2008; emphasis added. See also Victoria Jaggard, "At Ten, Dark Energy 'Most Profound Problem' in Physics," *National Geographic News*, May 16, 2008.

16. Turner cited in Richard Panek, "Out There," *New York Times*, March 11, 2007.

17. Schlegel cited in Panek, "Out There."

18. See NASA, "Dark Energy, Dark Matter," *Nasa Science: Astrophysics*, http://nasascience.nasa.gov/astrophysics/what-is-dark-energy.

19. As Rüdiger Vaas rather colorfully puts it, "If you imagine the universe as a cosmic cappuccino, the coffee stands for dark energy, the milk for dark matter, both of which we know almost nothing about; only the powdered chocolate would be what we are familiar with, namely ordinary matter made of protons, neutrons, electrons, et cetera." Rüdiger Vaas, "Dark Energy and Life's Ultimate Future," in *The Future of Life and the Future of Our Civilization*, ed. Vladimir Burdyuzha (Dordrecht: Springer, 2006), http://philsci-archive.pitt.edu/archive/00003271.

20. Krauss cited in Panek, "Out There."

21. These four are electromagnetism, gravity, the strong nuclear force, and the weak nuclear force. On the incompatibility of general relativity and quantum theory, and on the possibility that string theory might reconcile them, see Brian Greene, *The Elegant Universe: Superstrings, Hidden Dimensions, and the Quest for the Ultimate Theory* (New York: W. W. Norton, 1999). For a more critical view of string theory that nevertheless holds out hope that it might, in conversation with quantum loop gravity, become what it purports to be, see Lee Smolin, *Three Roads to Quantum Gravity* (New York: Basic Books, 2001).

22. See the 2003 report of the High Energy Physics Advisory Panel, "Quantum Universe: The Revolution in 21st Century Particle Physics," *Scribd.com*, http://www.scribd.com/doc/7778116/Quantum-UNIVERSE.

23. Perlmutter cited in Panek, "Out There."

24. Dennis Overbye, "Dark, Perhaps Forever," *New York Times*, June 3, 2008. On the possibility that we inhabit a multiverse, see Martin Rees, *Our Cosmic Habitat* (Princeton, N.J.: Princeton University Press, 2001).

25. Robert Caldwell of Dartmouth College, cited in Cowen, "Embracing."

26. Ibid.

27. Overbye, "Dark, Perhaps Forever."

28. The "Cappadocians" are Basil of Caesarea, Gregory Nazianzen, and Gregory of Nyssa. The distinction between essence and energy can be found in Basil in particular, who writes, "We say that we know the greatness of God, His power, His wisdom, His goodness, His providence over us, and the justness of His judgment, but not His very essence. . . . The energies are various and the essence simple, but we say that we know our God from His energies, but do not undertake to approach near to His essence. His energies come down to us, but His essence remains beyond our reach." Basil, *Epistle 234*, cited in David Bradshaw, "The Concept of Divine Energies," *Philosophy, Theology* 18, no. 1 (2006): 107. Gregory of Palamas confirms that "to say that the divine nature is communicable not in itself but through its energy, is to remain within the bounds of right devotion." Palamas in Vladimir Lossky, *The Mystical Theology of the Eastern Church* (London: James Clarke and Co., Ltd., 1957), 70.

29. According to Palamas, "God is called light not according to His essence, but according to His energy." Palamas in Vladimir Lossky, *In the Image and Likeness of God*, ed. John H. Erikson and Thomas E. Bird (Crestwood, N.Y.: St. Vladimir's Seminary Press, 1974), 41.

30. 2 Peter 1:4 in Lossky, *Mystical Theology*.

31. Lossky, *Mystical Theology*, 70.

32. Ibid., 74.

33. Gregory Palamas, *The Triads*, ed. John Meyendorff, trans. Nicholas Gendle, Classics of Western Spirituality (Mahwah, N.J.: Paulist Press, 1983), 96.

34. Lossky, *Mystical Theology*, 82.

35. Palamas, *Triads*, 96.

36. "Since it is the Cause of all beings, we should posit and ascribe to it all the affirmations we make in regard to beings, and, more appropriately, we should negate all these affirmations, since it surpasses all being." Pseudo-Dionysius, "The Mystical Theology," in *The Complete Works*, ed. Colm Luibheid, *Classics of Western Spirituality* (New York: Paulist Press, 1987), 100B.

37. Ibid., 1025B.

38. See Palamas, *Triads*, 80. Cf. Lossky, *Image*, 42.

39. See Palamas, *Triads*, 80. See also Lossky, who insists rather bafflingly that "The theology of darkness . . . will give way to a theology of the uncreated light. . . .The darkness of Mount Sinai will be changed into the light of Mount Tabor, in which Moses at last was able to see the glorious face of God incarnate." Lossky, *Image*, 42.

40. Rowan Williams has already done much of this work, although it is in defense of Thomistic orthodoxy, which this reflection finds equally limiting. See Rowan Williams, "The Philosophical Structure of Palamism," *Eastern Churches Review* 9 (1977): 27–44.

41. Matthew 17:5.

42. This term can be found throughout the *Divine Names* and in *The Mystical Theology* at 1048A. Palamas admits that Dionysius includes "*ousia*" among the divine names, but glosses over it quickly, lest it threaten his distinction. Palamas, *Triads*, 98.

43. See Bradshaw, "Concept of Divine Energies," 99.

44. "In this working God and I are one; he is working and I am becoming. The fire changes anything into itself that is put into it and this takes on fire's own nature. The wood does not change the fire into itself, but the fire changes the wood into itself." Meister Eckhart, "Sermon 6: Justi Vivent in Aeternum," in *The Essential Sermons, Commentaries, Treatise, and Defense*, ed. Edmund Colledge and Bernard McGinn, Classics of Western Spirituality (New York: Paulist Press, 1981), 189.

45. Jacques Derrida, "Différance," in *Deconstruction in Context*, ed. Mark C. Taylor (Chicago: University of Chicago Press, 1986), 399. See also Derrida, "How to Avoid Speaking: Denials," in *Derrida and Negative Theology*, ed. Harold Coward and Toby Foshay (Albany: State University of New York Press, 1992). For a comparison between Derrida and Dionysius on the question of teleology, see Mary-Jane Rubenstein, "Dionysius, Derrida, and the Problem of Ontotheology," *Modern Theology* 24, no. 2 (October, 2008): 725–42.

46. Alan H. Guth, *The Inflationary Universe: The Quest for a New Theory of Cosmic Origins* (New York: Helix Books, 1997). See also Alan H. Guth and David I. Kaiser, "Inflationary Cosmology: Exploring the Universe from the Smallest to the Largest Scales," *Science* 307 (2005).

47. See Steinhardt and Turok, *Endless Universe*, 9–10. Subsequent references will be cited in the main text.

48. The WMAP results, released in 2003, have confirmed that the universe is flat. This means that the density of the universe, designated by the Greek letter Omega ($\Omega$) is equal to 1. (If $\Omega$ were greater than one, the universe would be spherical; if it were less than one, the universe would be saddle-shaped.) At this moment, matter (both dark and light) makes up a bit less than 30 percent of the cosmos, while dark energy, which is expressed by Einstein's lambda ($\lambda$), occupies a bit more than 70 percent. Expressed in terms of the overall density of the universe, $\Omega_m = .27$ and $\Omega_\lambda = .73$. So the equation confirms the flatness of the universe: $\Omega_m + \Omega_\lambda = 1$. As time goes on, the lambda value will increase further, while the m-value will decrease proportionally.

49. This "flash" is said to have lasted between 10–36 seconds and 10–34 seconds ATB (After the Bang). See Greene, *Elegant Universe*, 356.

50. On the "big rip," see Robert Caldwell, Marc Kamionkowski, and Nevin N. Weinberg, "Phantom Energy and Cosmic Doomsday," *arXiv* (February 23, 2003). On the "big whimper," see Vaas, "Dark Energy and Life's Ultimate Future."

51. The possibility of a "Phoenix Universe" was posited in the 1920s by none other than Georges Lemaître; see Jean-Luc Lehners and Paul J. Steinhardt, "Dark Energy and the Return of the Phoenix Universe," *arXiv* (December 17, 2008). For reports of the resurrection of this hypothesis post-WMAP, see "Before the Big Bang," *BBC News*, April 10, 2001; Robert Roy Britt, "Brane-Storm Challenges Part of Big Bang Theory," *Space.com*, April 18, 2001; Alan Boyle, "Questioning the Big Bang," *MSNBC*, April 25, 2002; Richard Stengner, "Report: Universe Began by Colliding with Another," *CNN*, April 13, 2001.

52. "This is my reply to anyone who asks: 'What was God doing before he made heaven and earth?' My reply is not that which someone is said to have given as a joke to evade the force of the question. He said: 'He was preparing hells for people who inquire into profundities.'" Augustine of Hippo, *Confessions*, trans. Owen Chadwick (Oxford: Oxford University Press, 1991), 11.12.14.

53. Pius XII, "Modern Science and the Existence of God," *Catholic Mind* 49 (1952).

54. This is both under- and overstated. A number of theorists have gone to great lengths to insist the universe was created out of nothing (see Alex

Vilenkin, *Many Worlds in One: The Search for Other Universes* [New York: Hill and Wang, 2006] and Stephen Hawking and Leonard Mlodinow, *The Grand Design* [New York: Bantam, 2010]), while others have devised ways to avoid the ex nihilo (see Greene, *Elegant Universe*, 358; and Lee Smolin, *The Life of the Cosmos* [New York: Oxford University Press, 1997]).

55. In *Face of the Deep*, Catherine Keller unravels the complicated series of denials involved in constructing and defending the doctrine of *creatio ex nihilo*, when Genesis 1:2 seems to indicate that God created out of the deep of *tehom*. Keller, *Face of the Deep*, chapter 1.

56. Andrei Linde has contested the cyclical scenario's claim to provide an alternative to the inflationary hypothesis and offered the possibility of "eternal inflation" as a way to solve the "singularity problem" (that is, the reliance upon a moment of infinite temperature and pressure). See Andrei Linde, "Inflationary Theory Versus Ekpyrotic/Cyclic Scenario: A Talk at Stephen Hawking's 60th Birthday Conference, Cambridge University, Jan. 2002," *arXiv* (May 24, 2002). For a substantial revision of the hypothesis into a "New Ekpyrotic Cosmology" that "avoids any big crunch singularity," see Evgeny I. Buchbinder, Justin Khoury, and Burt A. Ovrut, "New Ekpyrotic Cosmology," *Physical Review D* 76, no. 12 (November 22, 2007). For another counter-argument, see Renata Kallosh et al., "The New Ekpyrotic Ghost," *arXiv* (March 26, 2008). And for a layperson's guide to the controversy, see Jon Cartwright, "Ekpyrotic Cosmology Resurfaces," *Physics World*, January 15, 2008.

57. Because it changes density, dark energy along this model is not Einstein's cosmological constant, but rather a dynamic substance called "quintessence." See Kirshner, *Extravagant Universe*, 358. For a longer discussion of quintessence, see Robert Caldwell and Paul J. Steinhardt, "Quintessence," *Physics World*, November 1, 2000.

58. See Paul J. Steinhardt, "A Brief Introduction to the Ekpyrotic Universe," http://wwwphy.princeton.edu/~steinh/npr.

59. Cicero, *On the Nature of the Gods*, trans. P. G. Walsh, Oxford World's Classics (New York: Oxford University Press, 2008), 2.118; emphasis added. Cited in Steinhardt and Turok, *Endless Universe*, 170.

60. Steinhardt and Turok address the cyclical-Hindu confluence in a remarkable passage that maps the life of Brahma mathematically. According to Hindu cosmology, the cycles of the universe are a day, a year, and a life of Brahma. A day lasts a *kalpa*, 8.64 billion years, which is almost exactly the length of the "matter" phase after the big bang. A year (8.64 × 360) lasts 3.11 trillion years, which is the length of one cycle in the cyclical model. The life of Brahma is 100 of these "years," and then the universe is said to rest. But then there are more brahmas to come (see Steinhardt and Turok, *Endless Universe*, 170).

61. M-theory posits eleven dimensions of space-time. See Greene, *Elegant Universe*, 283–319. No one seems to know what the "M" stands for; some have proposed "membrane," "matrix," "Mother (of All Theories)," or "mystery."

62. Each of these branes has three extended spatial dimensions, one dimension of time, and six imperceptible "compactified" dimensions in the form of a "Calabi-Yau manifold." The eleventh dimension runs between the two branes. See Lisa Randall, *Warped Passages: Unraveling the Mysteries of the Universe's Hidden Dimensions* (New York: Harper Collins, 2005), 330–32.

63. In most models, gravity is the only force that can travel across the "bulk" between branes. In Steinhardt and Turok's version, however, dark energy does the work both of attraction and of repulsion.

64. As is undoubtedly evident, it is not within my capabilities to judge the scientific validity of either of these theories. The most I hope to do in these pages is to throw one or two astrophysical logs on the theological fire. And while it is entirely possible that *neither* of these models is useful for theological reflection, I would suggest that the inflationary model, while it might end up to be "true," is, theologically speaking, not so much of a nonstarter as it is a dead end. After all, the Church has sustained itself for centuries on the ex nihilo, but only by maintaining a careful calibration between the explosive and attractive powers of God. Should it be the case that the former outruns the latter, and creation has happened once and for all, then "human participation in the divine energies" would mean little more than living into the runaway repulsion of the spheres. What participation might mean in a cyclical model is suggested in faltering and incomplete ways below.

65. Aristotle Papanikolaou, "Divine Energies or Divine Personhood: Vladimir Lossky and John Zizioulas on Conceiving the Transcendent and Immanent God," *Modern Theology* 19, no. 3 (July 2003): 359.

66. 1 Corinthians 12:4–11. For a full discussion of the role of *energein* and *energeia* in this passage, see Bradshaw, "Concept of Divine Energies," 105.

67. For a concrete account of such participation, see Charlene P. E. Burns, "Altruism in Nature as Manifestation of Divine Energeia," *Zygon* 41, no. 1 (March 2006): 125–28.

68. See David Bradshaw, *Aristotle East and West: Metaphysics and the Division of Christendom* (Cambridge: Cambridge University Press, 2004), 119–52; and Bradshaw, "The Divine Energies in the New Testament," *St. Vladimir's Theological Quarterly* 50 (2006): 189–223.

69. Bradshaw, "Concept of Divine Energies," 104; emphasis added.

70. Ibid., 118.

71. Meister Eckhart, "Sermon 52: Beati Paupers Spiritu, Quoniam Ipsorum Est Regnum Caelorum," in *The Essential Sermons, Commentaries, Treatise,*

*and Defense*, ed. Edmund Colledge and Bernard McGinn (New York: Paulist Press, 1981), 202.

72. Williams, "Philosophical Structure of Palamism," 37.

73. Ibid., 40–41.

74. Ecclesiastes 12:7, Steinhardt and Turok, *Endless Universe*, 225.

### 3. SOLAR ENERGY: THEOPHANY AND THE THEOPOETICS OF LIGHT IN GREGORY OF NYSSA
### *T. Wilson Dickinson*

1. This term has a number of meanings and uses, but for the sake of this essay, it will be primarily considered through the Cappadocian deployment of *energeia* as "activities" (as different from being [*ousia*] or nature [*physis*]), and secondarily in the Aristotelian and scholastic understanding of the term as "actuality" (which is aligned with being [*ousia*]).

2. See, for example, Martin Heidegger, "The Age of World Picture," in *Off the Beaten Track* (New York: Cambridge University Press, 2002), 73–85.

3. See Andrea Nightingale, *The Spectacles of Truth in Classical Greek Philosophy*: Theoria *in its Cultural Context* (Cambridge: Cambridge University Press, 2004); or Pierre Hadot, *What Is Ancient Philosophy?* (Cambridge: Harvard University Press, 2002).

4. Martin Heidegger, "The Question Concerning Technology," in *Basic Writings* (San Francisco: Harper Collins, 1993), 320.

5. Pierre Hadot, *The Veil of Isis: An Essay on the History of the Idea of Nature* (Cambridge: Belknap Press, 2006), 96.

6. Ibid., 35.

7. Ibid., 96.

8. Ibid., 215.

9. In appealing to Orpheus and Prometheus above, I am not intending to directly follow what seems to be a rather common line in eco-theology, which proposes a recovery of repressed Pagan elements in Christian and Western traditions that have a positive account of creation (see, for example, the work of Mark I. Wallace). Though I am bit skeptical of the pragmatic viability of such a strategy, I do not want to be antagonistic to such efforts. I am, instead, following a different reading of "Hellenic" Christian voices.

10. Aristotle, *De interpretation* 16a/Aristotle, *The Complete Works of Aristotle*, Vol. 1 (Princeton, N.J.: Princeton University Press, 1985), 25. Cited in Jacques Derrida, *Of Grammatology* (Baltimore: Johns Hopkins University Press, 1992), 11.

11. Wisdom 13:5.

12. Gregorius Nyssenus *contra Eunomium* [hereafter, Gr. Nyss. *Eun.*] 2.150/Gregory of Nyssa, "The Second Book Against Eunomius," trans.

Stuart George Hall, in *Gregory of Nyssa:* Contra Eunomium II: *An English Version with Supporting Studies, Proceedings of the 10th International Colloquium on Gregory of Nyssa (Olomouc, September 15–18, 2004)*, ed. Lenka Karfíková, Scot Douglass, and Johannes Zachhuber, Supplements to Vigiliae Christianae (Leiden: Brill, 2007), 91.

13. Both Scot Douglass and Alden Mosshammer emphasize this dialogical character of Gregory's "dialectic." My reading of Gregory here draws heavily from Douglass's study *Theology of the Gap: Cappadocian Language Theory and the Trinitarian Controversy* (New York: Peter Lang, 2005), and, to a lesser extent, Mosshammer's essay "Disclosing but Not Disclosed: Gregory of Nyssa as Deconstructionist," in *Studien zu Gregor von Nyssa und der christlichen Spatantike*, ed. Huberton Drobner and Christoph Klock (New York: E. J. Brill, 1990), 99–123.

14. In this regard Gregory frequently laments having to directly engage Eunomius's positions and struggles with how to proceed without falling into the absurd. At one point he declares exhaustion from this "pointless shadowboxing" that is only due to engaging Eunomius in his "logical trick over unbegottenness." Gr. Nyss. *Eun.* 2.172–73/trans. Hall, Contra Eunomium II, 96. Furthermore, one could argue that some of Gregory's more metaphysical statements are part of a rhetorical deployment of irony. In the midst of his argument against God using "literal words" and therefore following his argument for the language of nature, Gregory writes, "You will have seized a great tyranny, if you forcefully obtain for yourself this right, so that the very things you ban for others you yourself have the right to do, and the things you claim the right to attempt, you exclude other people from." Gr. Nyss. *Eun.* 2.311/trans. Hall, Contra Eunomium II, 128. Perhaps this is not a contradictory misstep but a clue to the ironic flourish of his own discourse. While I would hesitate to ascribe a Kierkegaardian plurivocity and playfulness to Gregory's authorial voice here, Mark Hart has argued for the Gregory's use of irony in other texts. See Mark Hart, "Gregory of Nyssa's Ironic Praise of Celibate Life," *Heythorp Journal* 33 (1992): 1–19.

15. For a critical reading of modern interpreters and their preoccupation with the trinity, see Morwenna Ludlow, *Gregory of Nyssa: Ancient and (Post)modern* (New York: Oxford University Press, 2007), 13–94. Elsewhere, Ludlow questions the degree to which Gregory is even systematic in these more doctrinal works. She writes, "Like a jazz pianist improvising, Gregory wanders: sometimes he appears to have lost the plot, but he always returns to the same key. This does not always make for an argument which is systematically presented and easy to follow: it is more impressionistic than systematic, more poetic than scientific." "Divinity Infinity and Eschatology," in Karfíková, *Gregory of Nyssa*, 225.

16. See Sarah Coakley, "Re-Thinking Gregory of Nyssa: Introduction– Gender, Trinitarian Analogies, and the Pedagogy of *The Song*," *Modern Theology* 18 (4 October 2003): 436.

17. Gregory's curriculum need not be read as reaching its high point in the speculative moments of *Against Eunomius*, but following the lead of Origen, he instead emphasizes spiritual exercises, particularly that of interpreting Scripture. He outlines an itinerary in which the "theologian" or "philosopher" learns ethics from Proverbs, physics from Ecclesiastes, and enoptics (*theoretikos*) from the Song of Songs. See Origenes *comm. in Cant.* Prologue 3; Gr. Nyss. *homiliae in Eccl.* 1; Gr. Nyss. *homiliae in Cant.* 1.

18. Gr. Nyss. *Eun.* 2.145/trans. Hall, Contra Eunomium II, 90.

19. Gr. Nyss. *Eun.* 2.304/trans. Hall, Contra Eunomium II, 127.

20. See, for example, Anthony Meredith, *Gregory of Nyssa* (New York: Routledge, 1999), 14.

21. Matthew 5:8.

22. He cites John 1:18, Exodus 33:20, and 1 Timothy 6:16.

23. Gr. Nyss. *de beatitudinibus* 6.1/Gregory of Nyssa, "Sermon 6," trans. Stuart George Hall, in *Gregory of Nyssa: Homilies on the Beatitudes: An English Version with Supporting Studies*, ed. Hubertus R. Drobner and Alberto Viciano (Boston: Brill, 2000), 66.

24. Matthew 14:28–32.

25. Gr. Nyss. *de beatitudinibus* 6.1/trans. Hall, *Homilies*, 66, 67.

26. Ibid., 66.

27. Gr. Nyss. *de beatitudinibus* 6.3.

28. Gr. Nyss. *de beatitudinibus* 6.4/trans. Hall, *Homilies* 69.

29. Ibid.

30. Ibid., 71.

31. Epistemological and metaphysical terms (*metaphysical fact, knowing subject*) are intentionally blended and employed here because the scholarly treatment of "Christian Neo-Platonism" seems to be derived largely from modern categories. The theoretical position that scholars often assume for themselves and the manner in which they treat their "textual objects" seem to entirely assume this frame even if they sometimes question the difference in content.

32. Even if one were to remain limited to metaphysics, this shift in thinking about *energeia* is still rather profound. Gregory's division between *energeia* and *ousia* is radically different from the typical Aristotelian and Scholastic schema in which God's being (*ousia*) is identified with God's actuality (*energeia*). The Scholastic schema contrasts actuality (*energeia*) with potency (*dunamis*), while Gregory uses the same terms but in an entirely different way, as he identifies activity (*energeia*) with power (*dunamis*) and differentiates them

from being (*ousia*). Accordingly, while Gregory's thought does not provide a direct response to the schematic of the former, this terminological shift does seem to lead to a radically different understanding about what one can know about God and, even more, the status of thought itself. Simply put, thinking about *energeia* in terms of activities or energies, rather than as that which is fully actual, seems to allow for an even greater sense in which that which is dynamic, created, and incomplete may be transformed and redeemed. What I am arguing for here, however, is that the radical character of the distinction between God's being (*ousia*) and activities (*energeia*) does not merely have a theoretical import, in which one can resituate the fundamental grammar of thinking by considering the particular alignments of the essential terms. For a general account of this difference in terminology, see Duncan Reid, *Energies of the Spirit* (Atlanta: Scholars Press, 1997), 7–26. In following this rather traditional story I do not suppose that this is the final or only story on Aristotle or the Scholastics.

33. Derrida, *Of Grammatology*, 18; emphasis added.

34. John D. Caputo distinguishes between the two, writing, "a poetics is an evocative discourse that articulates the event, while a logic is a normative discourse governing entities (real or possible), which can or do instantiate its proposition." *The Weakness of God: A Theology of the Event* (Bloomington: Indiana University Press, 2006), 103. Caputo also echoes this shift in thinking toward transformation, as he writes: "In a *poetics* of the impossible, we mean to pose the possibility of something life transforming, not to report how an omnipotent being intervenes upon nature's regularities and bends them to its infinite will" (104).

35. Ibid., 104.

36. Friedrich Nietzsche, *The Gay Science* (New York: Vintage Books, 1974), 315, 316.

37. Accordingly, one would have to couple the numerous instances in Gregory's "ascetic writings" that speak of cultivating the control that comes from *apatheia*, with the aforementioned admission that the complexities of our soul exceed our understanding.

38. Gr. Nyss. *Eun.* 2.79–80/trans. Hall, Contra Eunomium II, 77.

39. Matthew 11:16.

40. Gr. Nyss. *Eun.* 2.419/trans. Hall, Contra Eunomium II, 153.

41. Gr. Nyss. *Eun.* 2.81/trans. Hall, Contra Eunomium II, 77, 78.

42. Gr. Nyss. *de vita Mosis* [hereafter, Gr. Nyss. *v. Mos.*] 2.24/Gregory of Nyssa, *The Life of Moses*, trans., Abraham J. Malherbe and Everett Ferguson (New York: Paulist Press, 1978), 60. Furthermore, Gregory's definition of truth in this section is rendered by the translators as "the sure apprehension of real Being."

43. Gr. Nyss. *v. Mos.* 2.163/trans. Malherbe and Ferguson, *Life of Moses*, 95.

44. Gr. Nyss. *v. Mos.* 2.166/trans. Malherbe and Ferguson, *Life of Moses*, 96.

45. Gr. Nyss. *v. Mos.* 2.170–83.

46. Douglass, *Theology of the Gap*, 133, 134. In particular he is citing Exodus 40:34–35, and 2 Chronicles 6:14–21, 41.

47. Gr. Nyss. *v. Mos.* 2.170–83.

48. Gr. Nyss. *v. Mos.* 2.176–78.

49. Douglass, *Theology of the Gap*, 211.

50. Gr. Nyss. *v. Mos.* 2.184/trans. Malherbe and Ferguson, *Life of Moses*, 101.

51. Gr. Nyss. *v. Mos.* 2.200.

52. Gr. Nyss. *v. Mos.* 2.191/trans. Malherbe and Ferguson, *Life of Moses*, 103, 104.

53. Exodus 33:23.

54. Gr. Nyss. *v. Mos.* 2.51, 252.

55. Gr. Nyss. *v. Mos.* 2.253/trans. Malherbe and Ferguson, *Life of Moses*, 120.

56. Gr. Nyss. *Eun.* 2.183/trans. Hall, Contra Eunomium II, 98.

57. Gr. Nyss. *Eun.* 2.188/trans. Hall, Contra Eunomium II, 99.

## 4. BEYOND HEAT: ENERGY FOR LIFE
### *Clayton Crockett*

1. Philip Goodchild, *Theology of Money* (London: SCM Press, 2007), 261–62.

2. Ibid., 243.

3. See R. McNeill Alexander, *Energy for Animal Life* (Oxford: Oxford University Press, 1999), 5.

4. Ibid., 24.

5. J. C. Heesterman, *The Broken World of Sacrifice: An Essay in Ancient Indian Ritual* (Chicago: University of Chicago Press, 1993), 23.

6. I owe this term to Kevin Mequet.

7. See Thom Hartmann, *The Last Hours of Ancient Sunlight: The Fate of the World and What We Can Do before It's Too Late* (New York: Three Rivers Press, 2004).

8. See Kenneth Deffeyes, *Beyond Oil: The View from Hubbert's Peak* (New York: Hill and Wang, 2005), 29–30, 40.

9. For current production figures, see The Oil Drum (www.theoildrum .com). Keep in mind that the category Total Liquids includes other components including liquid natural gas; the best measure of oil extraction and pro-

duction is Crude + Condensates. The current (September 2009) EIA (Energy Information Administration, a function of the U.S. Department of Energy) estimates the highest month total (in millions of barrels per day) for both as July 2008 (http://www.theoildrum.com/node/5521#more).

10. George Monbiot, *Heat: How to Stop the Planet from Burning* (Cambridge, Mass.: South End Press, 2007), 6.

11. Ibid., 13.

12. See Herman E. Daly, *Beyond Growth: The Economics of Sustainable Development* (Boston: Beacon Press, 1996).

13. Roger Penrose, *The Emperor's New Mind: Concerning Computers, Minds, and the Laws of Physics* (Oxford: Oxford University Press, 1989), 186.

14. See Albert Einstein, *Relativity: The Special and the General Theory* (New York: Three Rivers Press, 1961), 24.

15. Einstein expresses the basic idea of general relativity in terms of Gaussian coordinate systems: "all Gaussian coordinate systems are essentially equivalent for the formulation of the general laws of nature." Einstein, *Relativity*, 108. What is a Gaussian coordinate system? In simple terms it means a definite but arbitrary coordination of points along a continuum, conditioned by the velocity of light. That is, there is no ultimate or objective reference frame or bodies within a reference frame that would allow one to plot coordinates to transfer from one system to another; we dispense with a body of reference to anchor the system and seek, rather, the rules of transformation of coordinate systems. These rules of transformation are the rules of energy conversion.

16. Werner Heisenberg, *Encounters with Einstein: And Other Essays on People, Places, and Particles* (Princeton, N.J.: Princeton University Press, 1989), 35.

17. See Penrose, *Emperor's New Mind*, 302–14.

18. Gilles Deleuze, *Difference and Repetition*, trans. Paul Patton (New York: Columbia University Press, 1994), 229.

19. Ibid., 225.

20. See Kevin Dobson Mequet, "Why It's Time to Develop a Fully Scalable NuMEgen Device Using the Feynmann/Gell-Mann Radioelectromagnetic Effect," 2009. Available online at http://www.scribd.com/doc/33017213/KDM-Proposal-NuMEgen-Vero3-Academic-20100605-Final.

21. We do not know whether or not Mercury has its own magnetic field, or whether it exhibits a magnetic-field effect that derives from its close proximity to the sun's powerful magnetic field.

22. See R. P. Feynman and M. Gell-Mann, "Theory of the Fermi Interaction," *Physical Review* 109, no. 1 (1958). See also the discussion in James

Gleick, *Genius: The Life and Science of Richard Feynman* (New York: Pantheon Books, 1992), 330–39.

23. Gilles Deleuze, "Immanence: A Life," in *Two Regimes of Madness: Texts and Interviews, 1975–1995*, trans. Ames Hodges and Mike Taormina (New York: Semiotext(e), 2006), 385–86.

24. On death-of-God theology and radical theology generally, see, among other works: Thomas J. J. Altizer, *The Gospel of Christian Atheism* (Philadelphia: Westminster Press, 1966); Gabriel Vahanian, *The Death of God: The Culture of Our Post-Christian Era* (New York: George Braziller, 1966); Thomas J. J Altizer and William Hamilton, *Radical Theology and the Death of God* (Indianapolis: Bobbs-Merrill, 1966); Carl Raschke et al., *Deconstruction and Theology* (New York: Crossroad, 1982); Mark C. Taylor, *Erring: A Postmodern A/theology* (Chicago: University of Chicago Press, 1984); Charles E. Winquist, *Desiring Theology* (Chicago: University of Chicago Press, 1995); and Mark C. Taylor, *After God* (Chicago: University of Chicago Press, 2007). For a more contemporary understanding of the death of God, see John D. Caputo and Gianni Vattimo, *After the Death of God*, ed. Jeffrey W. Robbins (New York: Columbia University Press, 2007). For two contemporary examples of radical theology, one more academic and one less so, see John D. Caputo, *The Weakness of God: A Theology of the Event* (Bloomington: Indiana University Press, 2006), and Peter Rollins, *How (Not) to Speak of God* (Brewster, Mass.: Paraclete Press, 2006).

25. See Leonard Lawlor, *The Implications of Immanence: Toward a New Concept of Life* (New York: Fordham University Press, 2006). Lawlor constructs a neo-vitalism drawing upon Deleuze, Derrida, Foucault, Merleau-Ponty, Heidegger, and Husserl.

5. EMERGENCE, ENERGY, AND OPENNESS: A VIABLE AGNOSTIC THEOLOGY
*Whitney Bauman*

1. Eric D. Schneider and Dorion Sagan, *Into the Cool: Energy Flow, Thermodynamics, and Life* (Chicago: University of Chicago Press, 2005). This chapter draws on a fuller discussion of a "viable agnostic theology," found in my book *Theology, Creation, and Environmental Ethics: From* Creatio ex Nihilo *to* Terra Nullius (New York: Routledge, 2009), especially chapters 6–8.

2. The metaphorical and material significance of this sentence gets blurry. Could it be that our "closed systems of thought" or our foundational thinking does indeed lead to entropy, death, and decay when forced onto evolving nature-cultures? This is what I argue below.

3. Schneider and Sagan, *Into the Cool*, 144.
4. Ibid., xvi.
5. Ibid., 81.

6. Ibid., 78. In the appendix of their book *Into the Cool*, Sagan and Schneider note that an organism's "organization does not come from nowhere but is paid for by an increase in the entropy of the greater 'universe.' The organized dissipative systems exist at the expense of greater disarray outside the system" (329).

7. I am thinking here of Mary Midgley's "model of the maps" in her *Science and Poetry* (London: Routledge, 2001). See especially part 2 of that book.

8. I discuss many of these authors below. After a very preliminary reading (I only began reading the book as this chapter was going to press), Mark C. Taylor's work *After God* (Chicago: University of Chicago Press, 2007) may be the closest to what I am arguing here. I look forward to further engagement with this text in future publications. The American Pragmatists, especially Rorty and Dewey, are in the background of my analysis of post-foundationalism as well. See, for example, John Dewey, *The Quest for Certainty: A Study of the Relation of Knowledge and Action* (1929; New York: G. P. Putnam's Sons, 1960); and Richard Rorty, *Philosophy and Social Hope* (London: Penguin, 1999).

9. For a good historical and philosophical analysis of emergence, see, for example, Mark A. Bedau and Paul Humphreys, *Emergence: Contemporary Readings in Philosophy and Science* (Boston: MIT Press, 2008).

10. Philip Clayton, "Conceptual Foundations of Emergence Theory," in *The Re-Emergence of Emergence: The Emergentist Hypothesis from Science to Religion*, ed. Philip Clayton and Paul Davies (Oxford: Oxford University Press, 2006), 2.

11. Ibid.

12. Arthur Peacocke, "Emergence, Mind, and Divine Action: The Hierarchy of the Sciences in Relation to the Human Mind-Brain-Body," in ibid., 259.

13. For a discussion of "ethical" and "epistemological" anthropocentrism, see Val Plumwood, *Environmental Culture: The Ecological Crisis of Reason* (London: Routledge, 2001), 167–95.

14. Gilles Deleuze and Félix Guattari seek to evoke such a metaphor in the "rhizome" and Bruno Latour in "the collective" (see discussion of both below), but I think it is just as important to evoke these embodied ways of becoming in the world using more sensuous metaphors. In other words, what does it feel like to embody "rhizomatic" thought of Deleuze and Guattari or the collective process that Latour describes?

15. See, for example, Catherine Keller, *God and Power: Counter-Apocalyptic Journeys* (Minneapolis: Fortress Press, 2005); and Dewey, *Quest for Certainty*.

16. René Descartes, *Meditations on First Philosophy III*, 42.

17. Bruno Latour, *The Politics of Nature: How to Bring the Sciences into Democracy*, trans. Catherine Porter (Boston: Harvard University Press, 2004), 147.

18. Ibid., 124.

19. Gordon Kaufman, *In the Beginning . . . Creativity* (Minneapolis: Fortress Press, 2004), 44.

20. Ibid., 45: "our *historicity* . . . is the most distinctive mark of our humanness." See also, Gordon Kaufman, *In Face of Mystery: A Constructive Theology* (Boston: Harvard University Press, 1995), 117.

21. Ronald Bogue, "A Thousand Ecologies," in *Deleuze/Guattari and Ecology*, ed. Bernd Herzogenrath (London: Palgrave, 2009), 49–50.

22. I follow a metaphysics of immanence from Bruno to Spinoza, and to the American Pragmatists, the theory of Emergence, and French "post" thinking such as is found in Gilles Deleuze and Félix Guattari's work.

23. "Nearly all [Christian groups] accepted the basic schema which elaborated a conception of God, and of God's Truth, as having independence and objectivity over against humanity." Gordon Kaufman, *An Essay on Theological Method* (Atlanta: Scholars Press of the American Academy of Religion, 1975), 28. See also: "Christians may no longer consider themselves to be authorized in what they say and do by God's special revelation." Kaufman, *In the Beginning*, 68. In fact, Kaufman also argues correctly that idealism and materialism are the same thing: "Materialisms themselves are, thus, at once products and examples of *spirit*, in the (empirical) sense in which I am using that word here." Kaufman, *In Face of Mystery*, 259.

24. Kaufman, *Essay*, 47.

25. Kaufman, *In Face of Mystery*, 67.

26. Catherine Keller, *Face of the Deep: A Theology of Becoming* (New York: Routledge, 2003). Also see, Bauman, *Theology, Creation*. Many of the ideas in this chapter began to take form in that book.

27. Kaufman, *Essay*, 40.

28. Ibid., 8.

29. Another, similar way to think of it is as what Lorraine Code describes (following Castoriadis) as "social imaginaries": "Imaginatively initiated counterpossibilities [that] interrogate the social structure to destabilize its pretensions to naturalness and wholeness, to initiate a new making (a *poesis*)." Lorraine Code, *Ecological Thinking: The Politics of Epistemic Location* (Oxford: Oxford University Press, 2006), 31.

30. Kaufman., *Essay*, 34.

31. Van A. Harvey's book is key for those who want to reread Feuerbach beyond the straw interpretation of "God as mere projection." Van A. Harvey,

*Feuerbach and the Interpretation of Religion* (Cambridge: Cambridge University Press, 1995).

32. Charles J. Sabatino, "Projection as Symbol: Rethinking Feuerbach's Criticism," *Encounter* 48, no. 2 (1987): 183.

33. Garrett Green, *Theology, Hermeneutics, and Imagination: The Crisis of Interpretation at the End of Modernity* (Cambridge: Cambridge University Press, 2000), 92. Furthermore, Green writes, "So the scholar of religion must say to Feuerbach: Yes, the imagination is indeed the source of religion, but No, religion is not thereby disqualified from the search for truth" (103).

34. Keller, *Face of the Deep*, 219–20.

35. Keller, *God and Power*, 118.

36. Val Plumwood discusses what a participatory democracy might look like in *Environmental Culture*, 93–96.

37. Rorty, *Philosophy and Social Hope*, 152.

38. See, for example, Gilles Deleuze and Félix Guattari, *A Thousand Plateaus: Capitalism and Schizophrenia*, trans. Brian Masumi (Minneapolis: University of Minnesota Press, 1987), 25: "Making a clean slate, starting or beginning again from ground zero, seeking a beginning or a foundation—all imply a false conception of voyage and movement . . . [rather we proceed] from the middle, through the middle, coming and going rather than starting and finishing."

39. To offer ground is not to make a foundational claim but to "give reasons, to cede turf, and to remember the shared earth that provides the one common ground in which all of our contexts nest." Catherine Keller and Anne Daniell, *Process and Difference: Between Cosmological and Poststructuralist Postmodernisms* (Albany: SUNY Press, 2002), 13. See also Keller's chapter in Catherine Keller and Laurel Kearns, eds., *EcoSpirit: Religions and Philosophies for the Earth* (New York: Fordham University Press, 2007), 63–76.

40. See, for example, Catherine Keller on John Cobb's concept of "the common good" in "Process and Chaosmos," in Keller and Daniell, *Process and Difference*, 55: "The university has organized knowledge/power according to standards of universality and objectivity that mask the special interests of race, class/economics, sex/gender, and species."

41. It is beyond these horizons that Kaufman's concept of "mystery" is helpful: "It is in terms of that which is beyond our understanding that we must, finally, understand our human language." Kaufman, *In Face of Mystery*, 6.

42. See, for example, J. Wentzel van Huyssteen, *The Shaping of Rationality: Toward Interdisciplinarity in Theology and Science* (Grand Rapids, Mich.: Eerdmans, 1999), 2: "Rationality is about responsibility: the responsibility to pursue clarity, intelligibility, and optimal understanding as ways to cope with ourselves and our world."

43. From within this theology as conversation model, reason, according to Kaufman, is "not . . . a kind of reservoir or bank from which our moral rules and principles can be withdrawn as needed; it is, rather, simply our critical capacity to discern, assess, and revise." Kaufman, *In Face of Mystery*, 193.

44. For a discussion of monological and dialogical ethical approaches, see Plumwood, *Environmental Culture*, 188–95.

45. Ludwig Feuerbach, *Lectures on the Essence of Religion*, trans. Ralph Manheim (New York: Harper and Row, 1967), 19.

46. Mark C. Taylor, *Erring: A Postmodern A/theology* (Chicago: University of Chicago Press, 1984), 25. In this book Taylor goes on to discuss the implications of the "death of the subject," "the end of history," and the "closure of the book" in ways that open up all of these concepts to continuing creation in a way I believe theology ought to open up to the "many otherings" of creation. He writes, "When it no longer seems necessary to reduce manyness to oneness [God, self, history, text] and to translate the equivocal as univocal, it becomes possible to give up the struggle for mastery and to take 'eternal delight' in 'The enigmatical Beauty' of each beautiful enigma" (176–77).

47. For a great analysis on the language of "dark energy" and "dark matter" as it relates to "light supremacy" and racism, see Barbara Holmes, *Race and the Cosmos: An Invitation to View the World Differently* (Harrisburg, Penn.: Trinity Press, 2002). This is, of course, a topic of much postcolonial analysis as well. E. G. Walter Mignolo, *The Darker Side of the Renaissance: Literacy, Territoriality and Colonialization* (Ann Arbor: University of Michigan Press, 2003); and Catherine Keller, Michael Nausner, and Mayra Rivera, eds., *Postcolonial Theologies: Divinity and Empire* (St. Louis, Mo.: Chalice Press, 2004).

48. John Llewellyn, *Margins of Religion: Between Kierkegaard and Derrida* (Bloomington: Indiana University Press, 2008).

6. ECOLOGICAL CIVILIZATIONS: OBSTACLES TO, AND PROSPECTS FOR,
RELIGIOUSLY INFORMED SUSTAINABILITY MOVEMENTS IN A
POST-AMERICAN WORLD
*Jay McDaniel*

1. See Mary Evelyn Tucker and John Grim, *Forum on Religion and Ecology* (New Haven, Conn.: Yale University Press, 2004), http://fore.research.yale .edu/religion/ (accessed August 5, 2009).

2. See Jay McDaniel, "In the Beginning Is the Listening," in *Ecology, Economy, and God*, ed. Darby Kathleen Ray (Minneapolis: Fortress Press, 2006), 26–42.

3. United States National Intelligence Council, *Global Trends, 2025: A Transformed World* (Washington, D.C.: U.S. Government Printing Office, 2008).

4. "Ecological Civilization," Opinion/Commentary (*China Daily*, Oct. 24, 2007), http://www.chinadaily.com.cn/opinion/2007-10/24/content_6201964.htm (accessed August 6, 2009).

### 7. "ONE MORE STITCH": RELATIONAL PRODUCTIVITY AND CREATIVE ENERGY
*Donna Bowman*

1. See, for example, Ray Kurzweil, *The Singularity Is Near: When Humans Transcend Biology* (New York: Viking, 2005).
2. See, for example, Bill Joy, "Why the Future Doesn't Need Us," *Wired* 8, no.4 (2004), http://www.wired.com/wired/archive/8.04/joy.html.
3. Max Chafkin, "The Customer Is the Company," *Inc.* June 2008, http://www.inc.com/magazine/20080601/the-customer-is-the-company.html.
4. Tim O'Reilly, "The Significance of Threadless.com," Nov. 20, 2006, http://radar.oreilly.com/archives/2006/11/the-significanc.html.
5. I first stated these ideas, in similar language, in a blog post titled "How the Web Changed Knitting (And Knitting Changed the World)," http://uniontrueheart.blogspot.com/2008/05/how-web-changed-knitting-and-knitting.html.

### 8. ENERGY, ECOLOGY, AND INTENSIVE ALLIANCE: BRINGING EARTH BACK TO HEAVEN
*Luke B. Higgins*

1. It does not seem to be an accident that the same term is used to describe the exchange of money, although an investigation of the fascinating linkages between capitalism and this techno-scientific rationality is outside the scope of this essay.
2. Bruno Latour, *Politics of Nature: How to Bring the Sciences into Democracy*, trans. Catherine Porter (Cambridge: Harvard University Press, 2004).
3. As we shall see shortly, I do not use the term *divinize* lightly. It will also become clear that I think *homo faber* as a definition of humanity is only problematic insofar as it assumes that *only* human beings possess this creative/constructive capacity.
4. Martin Heidegger coins this term to describe the particular approach to the natural world that develops within the mode of "enframing," or *gestell*. See Martin Heidegger, "The Question Concerning Technology," in *Basic Writings*, ed. David Farrell Krell (San Francisco: Harper, 1977). While Heideggers's critique of technology shares certain features with the Whiteheadian critique I am utilizing here, I believe the latter carries certain advantages—most significantly, a greater capacity to cooperate with science rather than pit philosophy against science.

5. I draw primarily from the following work: Ilya Prigogine and Isabelle Stenger, *Order Out of Chaos: Man's New Dialogue with Nature* (New York: Bantam Books, 1984). Whitehead's analysis of science is perhaps best articulated in his book *Science and the Modern World* (New York: Free Press, 1925).

6. Amory Lovins, *Soft Energy Paths: Towards a Durable Peace* (New York: Harper, 1977).

7. Janine M. Benyus, *Biomimicry: Innovation Inspired by Nature* (New York: Harper Perennial, 2002).

8. Michel Serres, *The Natural Contract*, trans. Elizabeth MacArthur and William Paulson (Ann Arbor: University of Michigan Press, 1995).

9. See in particular Val Plumwood's fascinating ecofeminist critique of Western, dualistic reasoning, *Environmental Culture: The Ecological Crisis of Reason* (London: Routledge, 2002).

10. See Martin Heidegger, *Being and Time*, trans. John MacQuarrie and Edward Robinson (New York: Harper Perennial, reprt. ed., 2008). Jacques Derrida critiqued what he saw as too neat a division between ontology and theology in Heidegger's work, but continued on with a critique of the "theological" tendencies of a certain kind of metaphysical thinking of "presence." Jacques Derrida, *Of Grammatology*, trans. Gayatri Spivak (Washington, D.C.: Johns Hopkins University Press, 1998). For Derrida's critique of Heidegger, see *Of Spirit: Heidegger and the Question*, trans. Geoffrey Bennington and Rachel Bowlby (Chicago: University of Chicago Press, 1987).

11. Carolyn Merchant, *Reinventing Eden: The Fate of Nature in Western Culture* (New York: Routledge, 2004).

12. This identification would, of course, centrally turn on the ancient concept of *logos*, a term that, in its time, opened up an interface between notions of creation, rationality, word/language, and a personal, divine agency that in the case of Christ was understood to take human form.

13. Prigogine and Stengers, *Order Out of Chaos*, 47.

14. Ibid., 49–50.

15. "At the origin of modern science, a "resonance" appears to have been set up between theological discourse and theoretical and experimental activity—a resonance that was no doubt likely to amplify and consolidate the claim that scientists were in the process of discovering the secret of the "great machines of the universe." Prigogine and Stengers, *Order Out of Chaos*, 46.

16. Ibid., 305.

17. Ibid., 50.

18. Ibid., 42.

19. Ibid., 51.

20. Ibid., 67.

21. Ibid., 76.

22. The more evangelical eco-theologies tend to rely most exclusively on the idea of obeying God's injunction to care for creation. See Steven Bouma-Prediger, *For the Beauty of the Earth: A Christian Vision for Creation Care* (Grand Rapids, Mich.: Baker Academic, 2001), and Calvin DeWitt, *Caring for Creation: Responsible Stewardship of God's Handiwork* (Grand Rapids, Mich.: Baker Academic, 1998).

23. Other fascinating explorations of this critique are offered by Bruno Latour in *Politics of Nature*. See also Jane Bennett, *The Re-Enchantment of the World: Secular Magic in a Rational Age* (Princeton, N.J.: Princeton University Press, 2001).

24. Alfred N. Whitehead, *Process and Reality, Corrected Edition*, ed. David Ray Griffin and Donald Sherburne (New York: Free Press, 1978), 94.

25. Ibid., 93.

26. Thus, process thought is better understood as affirming a doctrine of pan-experientialism rather than pan-psychism.

27. Whitehead, *Process and Reality*, 83; my emphasis.

28. Ibid., 111.

29. Ibid., 105.

30. The particular version of process theology I am advocating here places an emphasis on the immanent aspects of God's agency in the world. While the classic expositions of process theism certainly have a place for this immanence, it tends to get balanced with a more traditional understanding of transcendence, and thus God is still spoken of as an entity that can be encountered separately from the concrete relationships of our lives. See John Cobb and David Ray Griffin, *Process Theology: An Introductory Exposition* (Louisville, Ky.: Westminster John Knox, 1976) and Marjorie Suchocki, *God, Christ, Church: A Practical Guide to Process Theology* (New York: Crossroad, 1989). While my approach is no radical departure from Cobb, Griffin, and Suchocki, it cleaves a bit more consistently to Whitehead's belief that eternal objects, including their "envisagement" in God's primordial vision, have no realized value until they are ingressed within particular configurations of actual occasions.

31. Whitehead, *Process and Reality*, 254.

9. "GO BIG OR GO HOME": A CRITIQUE OF THE WESTERN CONCEPT
OF ENERGY/POWER AND A THEOLOGICAL ALTERNATIVE
*Oz Lorentzen*

1. Robert Crease writes, "the first technical definition of the word being provided by Aristotle. His definition was, however, different from the one that we use today. Every existing thing, he said, has an *energeia* that maintains it in being and is related to its end or function, or *telos*. He called a body's

potential or capacity for action its dynamis, and used *en-ergeia* to refer to the body being 'at work' en route to—or at—that telos" (http://physicsworld .com /cws/article/print/9233).

2. I propose, in brief, two main arguments for the materialist approach to energy being an error, one logical and the other ethical. The first is that it is a *reductio ad absurdum*, the fact that we have ended up with this untenable, "crisis" energy situation (not just the inevitable disappearance of "cheap fuel," but equally the environmental and ecological devastation) is an argument for an error in the beginning premise. Second, the materialist approach necessarily is a devaluing (a loss of values), since it removes any questions of intrinsic value/telos. As such, it is a blind guide. It can create systems by which the cosmos is known, engaged, manipulated, but it has *lost the ability to guide* this process. Nuclear energy, for example, the historic use of the atom bomb, seems to belie the "value-free" nature of such knowledge.

3. Berry's words are: "Because this earlier situation makes serious demands upon the human in return for the benefits given, the industrial age was invented to avoid the return due for the benefits received. The burdens imposed upon the human in its natural setting . . . established a situation unacceptable to an anthropocentric community with its *deep psychic resentment* against any such demands imposed upon it." Thomas Berry, "Creative Energy," in *Cross Currents* 37, no. 2–3 (1987): 179–86; quote on p. 183; my emphasis.

4. By "a slave to the systems . . .," I refer to the sort of work that Jacques Ellul has done on the impact of technology on society. See Jacques Ellul, *The Technological Society* (New York: Vintage, 1967).

5. Of course, since all theology is (at least in part) autobiographical, I hold this position for both rational and experiential reasons as well.

6. For this distinction I am indebted to a conversation with a colleague, Michael Pahl.

7. A theological strategy employed and outlined by Charles Winquist in *Desiring Theology* (Chicago: University of Chicago Press, 1995).

8. Biblical quotes are from the New American Standard Version.

9. This is a central command, given its place in the Decalogue, its function in the life and history of the Jewish people, and its reiteration in the Christian corpus in Hebrews.

10. I borrow the phrase "epistemology of the cross" and concept from Carl Holladay, *A Critical Introduction to the New Testament* (Nashville, Tenn.: Abingdon Press, 2005).

11. See Alain Badiou, *Saint Paul: The Foundation of Universalism*, trans. Ray Brassier (Stanford, Calif.: Stanford University Press, 2003). The rest of this section has resonances with Badiou's reading of Paul.

12. Another illustrative point comes from Gadamer's idea of the educated person in *Truth and Method*: not someone who has all the information at ready, and thus has the answer (in this case a human encyclopedia or computer), but someone who training/learning prepares to be ready to listen and question (a skilled but open mind, not a full mind). See David Blacker's related discussion of Gadamer in "Education as the Normative Dimension of Philosophical Hermeneutics," in *Philosophy of Education*, 1993 (http://www .ed.uiuc.edu/EPS/PES-Yearbook/93_docs/Blacker.HTM).

13. For instance, St. John of the Cross's concept of the "Dark Night of the Soul," where the human faculties are "darkened" so that the true light can be accessed, although this accessing does not lessen the darkness.

14. See Michael Jinkins, *An Invitation to Theology* (Westmount, Ill.: Intervarsity Press: 2006), for an elaboration of this point.

15. See Steven Bryan, "Power in the Pool: The Healing of the Man at Bethesda and Jesus' Violation of the *Sabbath* (Jn. 5:1–18)," in *Tyndale Bulletin* 54, no. 2 (2003): 7–22, for a discussion of this sort of thinking being a backdrop for the confrontation between Jesus and the Jewish leaders surrounding a healing on the Sabbath, the context for one of the Sabbath verses above.

16. C. S. Lewis makes this point in *The Abolition of Man* (New York: Harper One, 2001). He is referring here to the heirs of the medieval alchemy, for example, the quest for the philosopher's stone, etc.

17. Berry, "Creative Energy," 184.

18. A crisis not just because "cheap" energy is running out, but because of the devastating legacy of "cheap" energy.

19. "'Wu-wei' literally means 'in the absence of/without doing' and is often translated as 'doing nothing' or non-action.' It is important to realize, however, that wu-wei properly refers not to what is actually *happening* (or not happening) in the realm of observable action but rather to the *state of mind* of the actor. . . . It describes a state of personal harmony in which actions flow freely and instantly from one's spontaneous inclinations . . . and yet nonetheless perfectly accord with the dictates of the situation at hand. . . . As Jean-Francois Billeter describes it, wu-wei . . . represents a state of 'perfect knowledge of the reality of the situation, perfect efficaciousness and the *realization of a perfect economy of energy.*'" Edward Slingerland, "Effortless Action: The Chinese Spiritual Ideal of Wu-wei," *Journal of the Academy of Religion* 68, no. 2 (2000): 293–328, quoted text on 299–300; emphasis added. See also the comment by James Stine: "*wu-wei* suggests a kind of *kenosis* on humanity's part—that is the self-emptying, quietness, stillness by which a human being allows the Way to be (the Way!) through him or her. *The Tao never appears to pretense and its lifelessness. . . . Wei wu-wei* is to act without action through stillness and silence." Stine, "I Am the Way: Michael Polanyi's Taoism," *Zygon* 20, no. 1 (1985): 59–77, quote on p. 60; emphasis mine.

20. It is interesting to note that this perspective was seen by some as a potential contender for an explanation/understanding of the world at the turn of the nineteenth century. I refer to the idealism, spiritualism, growth of new religions, and so forth during that time. One could argue that the pragmatic success of the materialist view led to an almost complete eclipse of this view in the last one hundred years. The New Age religious/spirituality is a return to some of these perspectives. See Henri Bergson's essay for an example of a "scientific" sympathy of these sorts of phenomenon: "'Phantasms of the Living' and 'Psychical Research'" in *Mind Energy; Lectures and Essays*, trans. H. W. Carr (London: Macmillan and Co., 1920).

21. Genesis 1:28 ("God blessed them; and God said to them, "Be fruitful and multiply, and fill the earth, and subdue it; and rule over the fish of the sea and over the birds of the sky and over every living thing that moves on the earth") is called the creation covenant or creation mandate. This is the text many go to as a beginning point for a theological approach to the human life in the natural order.

22. Maybe utilizing something like the early work by Howard Odum, "Ecosystem, Energy, and Human Values," *Zygon* 12, no. 2 (1977): 109–33.

23. The idea of "flourishing" is a concept drawn from Aristotle's virtue ethics in the *Nicomachean Ethics*.

### 10. GOD IS GREEN; OR, A NEW THEOLOGY OF INDULGENCE
#### *Jeffrey W. Robbins*

1. Dietrich Bonhoeffer, *Ethics* (New York: Simon and Schuster, 1995), 66.

2. Ibid., 66–67.

3. Ibid., 70.

4. See Benjamin Barber, *Jihad vs. McWorld: Terrorism's Challenge to Democracy* (New York: Ballantine Books, 1995), esp. pp. 236–46.

5. Vine Deloria Jr., *God Is Red: A Native View of Religion* (New York: Grosset and Dunlap, 1973). Page number(s) for future references are in parentheses.

6. Incidentally, this reading of the 1960s counterculture is shared by Mark C. Taylor in his recent book *After God*. As Taylor writes: "Protests to the contrary notwithstanding, the counterculture was a *religious* movement. The culture wars and religious wars of the past two decades have demonized the sixties as a decadent period when individuals and society lost their way by slipping into the morass of relativism, which, critics argue, inevitably leads to nihilism. But this is a naïve and simplistic view. . . The religious language of redemption and salvation was translated into the language of personal fulfillment and social liberation. Though the terms changed, the questions remained the same as have been asked throughout the history of Christianity: Is liberation primarily personal or social? In other words, must one change

consciousness to change society or change society to change consciousness?" In Mark C. Taylor, *After God* (Chicago: University of Chicago Press, 2007), 249, 251.

7. For instance, see Bill McKibben, *The End of Nature* (New York: Random House, 2006); Denis Edwards, *Ecology at the Heart of Faith* (Maryknoll, N.Y.: Orbis Books, 2006); and Sallie McFague, *A New Climate for Theology: God, the World, and Global Warming* (Philadelphia: Fortress Press, 2008), among others.

8. Lisa Margonelli, *Oil on the Brain: Adventures from the Pump to the Pipeline* (New York: Doubleday, 2007), 103–04. Page number(s) for future references are in parentheses.

9. Kevin Phillips, *American Theocracy: The Peril and Politics of Radical Religion, Oil, and Borrowed Money in the 21st Century* (New York: Viking, 2006). Page numbers of future reference(s) are in parentheses.

10. Paul Roberts, *The End of Oil: On the Edge of a Perilous New World* (New York: Houghton Mifflin, 2004), 3.

11. This distinction between conservation and efficiency is an important one as evidenced by a recent cover story from Time magazine that makes the case that if the United States develops better energy efficiency, little or nothing needs to be changed in our ingrained habits or way of life. Conservation, on the other hand, requires sacrifice. See "Wasting our Watts," *Time* (January 12, 2009), 32–36.

12. As summarized by Richard Lacayo in his review for *Time* magazine. See "The Unholy Alliance," *Time* (March 19, 2006), http://www.time.com/time/printout/0,8816,1174713,00.html.

13. Mark C. Taylor, *Confidence Games: Money and Markets in a World without Redemption* (Chicago: University of Chicago Press, 2004), 6.

14. "Oil and the Dollar," *Wall Street Journal*, January 4, 2008, http://online.wsj.com/article/SB119941453085566759.html.

WHITNEY BAUMAN is assistant professor of science and religion at Florida International University in Miami. He is the author of *Theology, Creation, and Environmental Ethics: From* Creatio ex Nihilo *to* Terra Nullius (Routledge, 2009) and co-editor of *Grounding Religion: A Field Guide to the Study of Religion and Ecology* (Routledge, 2010). He serves as the co-chair of the Religion and Ecology Group of the American Academy of Religion and as book review editor for *Worldviews: Global Religions, Culture, and the Ecology.*

DONNA BOWMAN is associate dean and associate professor in the Honors College at the University of Central Arkansas. She holds a PhD in philosophical theology from the University of Virginia. Donna is the author of *The Divine Decision: A Process Doctrine of Election* (Westminster John Knox Press, 2002) and co-editor (with Jay McDaniel) of *Handbook of Process Theology* (Chalice Press, 2006).

CLAYTON CROCKETT is associate professor and director of religious studies at the University of Central Arkansas. He is the author of three books, including *Interstices of the Sublime: Theology and Psychoanalytic Theory* (Fordham University Press, 2007) and, most recently, *Radical Political Theology* (Columbia University Press, 2011).

T. WILSON DICKINSON is a doctoral candidate in the Department of Religion at Syracuse University and a visiting assistant professor of philosophy at Transylvania University. His work focuses on nineteenth- and twentieth-century continental philosophy and Christian theology, particularly as they address issues that arise from globalization. His dissertation is titled "Specters of Truth: Exercising Philosophy and Theology."

LUKE B. HIGGINS received his MDiv from the Pacific School of Religion in Berkeley, California, and is currently completing his PhD dissertation at Drew University. Drawing on the philosophies of life of Bergson, Deleuze, and Whitehead, his dissertation develops an interpretation of the kenotic incarnation of the Logos that can serve as the basis for an ecological, non-anthropocentric doctrine of creation. He is also an adjunct professor of philosophy at Rockland Community College in Suffern, New York, and is director of Christian education at a UCC church in Cedar Grove, New Jersey.

CATHERINE KELLER is professor of constructive theology in the Graduate Division of Religion and the Theological School of Drew University. She explores the overlaps of recent philosophical, feminist, and ecopolitical theologies in interaction with wider traditions of cosmology and mysticism. Recent books she has authored include *Face of the Deep: A Theology of Becoming; God and Power; and On the Mystery* (Routledge, 2003). She has co-edited multiple volumes for the Drew Transdisciplinary Theological Colloquium series, including most recently *Apophatic Bodies* (Fordham University Press, 2010) and *Polydoxy* (Routledge, 2011).

OZ LORENTZEN is currently the director at King's Fold Retreat and Renewal Centre, outside of Calgary, Alberta. He has an MA in religion and culture from Carleton University, Ottawa, and a PhD in religious studies from Syracuse University. Oz has been on faculty at SUNY-Morrisville and Prairie College in Alberta. His teaching and research interests lie in the area of faith and reason and, pedagogically, in the area of holistic knowledge and learning.

JAY MCDANIEL is director of the Steel Center for the Study of Religion and Philosophy at Hendrix College. His books include *With Roots and Wings: Christianity in an Age of Ecology* (Orbis Books, 1995) and *Handbook of Process Theology* (edited with Donna Bowman; Chalice Press, 2006). His research interests lie at the intersections of ecology, religion, continental philosophy, and culture studies.

JEFFREY W. ROBBINS's area of specialization is in continental philosophy of religion. He is associate professor of religion at Lebanon Valley College, where, in addition to teaching courses in religion, he is the director of the American Studies program and the college colloquium. He is the author of three books, including *Radical Democracy and Political Theology* (Columbia University Press, 2011), and editor of two others, including most recently *The Sleeping Giant Has Awoken: The New Politics of Religion in the United States* (Continuum, 2008). He is also the associate editor of the *Journal for Cultural and Religious Theory* and co-editor of the Columbia University Press book series Insurrections: Critical Studies in Religion, Politics, and Culture.

MARY-JANE RUBENSTEIN is assistant professor of religion at Wesleyan University, where she teaches in the areas of philosophy of religion and modern Christian thought. She is the author of *Strange Wonder: The Closure of Metaphysics and the Opening of Awe* (Columbia University Press, 2009) and of numerous articles on negative theology, Kierkegaard, Derrida, and the crisis over sex and gender in the worldwide Anglican Communion. She is currently writing a book on cosmologies of the multiverse.

Harvey, Van A., 181*n*31
Hawking, Stephen, 171*n*54
Hawkins, Richard, 163*n*9
*Heat: How to Stop the Planet from Burning,*
    178*n*10
Hebrews, 144, 187*n*9
Heesterman, J. C., 61, 177*n*5
Hegel, Georg, 35
Heidegger, Martin, 44, 126, 173*n*4,
    179*n*25, 184*n*4, 185*n*10
Heisenberg, Werner, 64, 178*n*16
Helweg-Larsen, Tim, 163*n*9
Heraclitus, 21
Higgins, Luke B., 6
Hitler, Adolf, 151
Holladay, Carl, 187*n*10
Holmes, Barbara, 183*n*47
Holmes, Tim, 163*n*9
Hooper, Dan, 28, 165*n*3, 166*n*6, 167*n*12
Horkheimer, Max, 74
*How (Not) to Speak of God*, 179*n*24
Hu Jintao, 92
Hubbert, M. K., 62
Hubble, Edwin, 26–27, 166*n*8
Humphreys, Paul, 180*n*9
Hunt, Christian, 163*n*9
Husserl, Edmund, 179*n*25
Huterer, Dragan, 166–67
Huyssteen, J. Wentzel van, 71, 82,
    182*n*42

*The Implications of Immanence: Toward a*
    *New Concept of Life*, 179*n*25
*In the Beginning . . . Creativity*, 181*n*19
*In Face of Mystery: A Constructive Theology,*
    181*n*20, 181*n*23, 181*n*25, 182*n*41,
    183*n*43
*In the Image and Likeness of God*, 168*n*29
*The Inflationary Universe: The Quest for*
    *a New Theory of Cosmic Origins,*
    170*n*46
*Into the Cool: Energy Flow, Thermodynam-*
    *ics, and Life*, 70, 179*n*1, 179*nn*3–5,
    180*n*6
*An Invitation to Theology*, 188*n*14
Isaac, 143
Isaiah, 141, 147

Jacob, 143
Jaggard, Victoria, 167*n*15
James, 56
Jesus, 25, 48, 59, 142–43, 188*n*15

*Jihad vs. McWorld: Terrorism's Challenge to*
    *Democracy*, 189*n*4
Jinkins, Michael, 188*n*14
Job, 24, 165*n*36
John, 56, 143–44, 175*n*22
John of the Cross, 188*n*13
Joy, Bill, 184*n*2
Joyce, James, 18

Kaiser, David I., 170*n*46
Kamionkowski, Marc, 170*n*50
Kant, Immanuel, 149
Kaufman, Gordon, 71, 73, 78,
    82, 181*nn*19–20, 181*nn*23–25,
    181*nn*27–28, 181*n*30, 182*n*41, 183*n*43
Kearns, Laurel, 17, 182*n*39
Keller, Catherine, 3–5, 71, 76, 80–81,
    90, 97, 108, 164*n*16, 165*n*34, 165*n*37,
    166*n*7, 171*n*55, 180*n*15, 181*n*26,
    182*nn*34–35, 182*nn*39–40, 183*n*47
Keshgegian, Flora, 1, 3
Khoury, Justin, 171*n*56
Kierkegaard, Søren, 35, 155, 174*n*14
King, Martin Luther Jr., 92
Kirshner, Robert P., 166*n*3, 171*n*9
Krauss, Lawrence, 29–30, 167*n*20
Kristeva, Julia, 89
Kurzweil, Ray, 107, 184*n*1

Lacan, Jacques, 89
Lacayo, Richard, 190*n*12
*The Last Hours of Ancient Sunlight: The*
    *Fate of the World and What We Can Do*
    *Before It's Too Late*, 177*n*7
Latour, Bruno, 19, 76–77, 122, 164*n*17,
    180*n*14, 181*n*17, 184*n*2, 186*n*23
Lawlor, Leonard, 179*n*25
*Lectures on the Essence of Religion*, 183*n*45
Lehners, Jean-Luc, 170*n*51
Lemaître, Georges, 27, 35, 166*n*5,
    170*n*51
Leucippus, 166*n*5
Levinas, Emanuel, 89
Lewis, C. S., 188*n*16
*The Life of the Cosmos*, 171*n*54
*The Life of Moses*, 54–56, 176*n*42,
    177*nn*43–45, 177*nn*47–48, 177*nn*50–
    52, 177*nn*54–57
*The Life and Science of Richard Feynman,*
    178*n*22
Linde, Andrei, 171*n*56
Livio, Mark, 30